PHILOPONUS

On Aristotle

Posterior Analytics 1.19-34

PHILOPONUS

On Aristotle Posterior Analytics 1.19-34

Translated by Owen Goldin
and Marije Martijn

B L O O M S B U R Y
LONDON • NEW DELHI • NEW YORK • SYDNEY

Bloomsbury Academic
An imprint of Bloomsbury Publishing Plc

50 Bedford Square
London
WC1B 3DP
UK

1385 Broadway
New York
NY 10018
USA

www.bloomsbury.com

Bloomsbury is a registered trade mark of Bloomsbury Publishing Plc

First published in 2012 by Bristol Classical Press
Paperback edition first published 2014

British Library Cataloguing-in-Publication Data
A catalogue record for this book is available from the British Library.

ISBN HB: 978-1-7809-3090-9
PB: 978-1-4725-5799-5
ePDF: 978-1-4725-0175-2

Library of Congress Cataloging-in-Publication Data
A catalog record for this book is available from the Library of Congress.

The present translations have been made possible by generous and imaginative funding from the following sources: the National Endowment for the Humanities, Division of Research Programs, an independent federal agency of the USA; the Leverhulme Trust; the British Academy; the Jowett Copyright Trustees; the Royal Society (UK); Centro Internazionale A. Beltrame di Storia dello Spazio e del Tempo (Padua); Mario Mignucci; Liverpool University; the Leventis Foundation; the Arts and Humanities Research Council; Gresham College; the Esmée Fairbairn Charitable Trust; the Henry Brown Trust; Mr and Mrs N. Egon; the Netherlands Organisation for Scientific Research (NWO/GW); the Ashdown Trust; Dr Victoria Solomonides, the Cultural Attaché of the Greek Embassy in London. The editor wishes to thank Jonathan Barnes, Orna Harari, Mariska Leunissen and Richard McKirahan for their comments, Sebastian Gertz for preparing the volume for press, and Deborah Blake, who has been the publisher responsible for every volume since the first.

Typeset by Ray Davies

Contents

Conventions

[...] Square brackets enclose words or phrases that have been added to the translation or the lemmata for purposes of clarity, as well as those portions of the lemmata which are not quoted by Philoponus.

<...> Angle brackets enclose conjectures relating to the Greek text, i.e. additions to the transmitted text deriving from parallel sources and editorial conjecture, and transposition of words or phrases. Accompanying notes provide further details.

(...) Round brackets, besides being used for ordinary parentheses, contain transliterated Greek words and Bekker page references to the Aristotelian text.

Introduction

This book is the third volume of a translation into English of a commentary on Aristotle's *Posterior Analytics* written by the late Neoplatonic commentator Philoponus (*c.* 490-*c.* 570). We are told initially that the commentary is 'based on meetings with Ammonius, son of Hermeias, with some of his own observations' (1,2-4). That is to say that while the commentary is in the main a record of the close reading of the *Posterior Analytics* offered in lectures by Philoponus' teacher Ammonius (*c.* 440-*c.* 520), Philoponus allows himself to include original observations and interpretations.[1] On occasion Philoponus identifies Ammonius as the source of an observation or interpretation. Otherwise, the reader may take it as probable, but not certain, that points of interpretation have Ammonius as their origin.

Included in this volume is Philoponus' commentary on 1.19-34. It begins with Philoponus' account of the long and complex argument of 1.19-23, to the effect that scientific demonstrations can neither be infinitely long nor infinitely extendible through the interposition of new middle terms. In commenting on this argument, Philoponus offers a distinction, familiar from the later scholastic tradition, between 'natural predications', in which a term referring to the ontological subject of the predicate serves as the grammatical subject, and 'unnatural predications', in which this is not so. Philoponus also here offers an interesting discussion of Aristotle's rejection of Platonic Forms. In 1.24 Aristotle argues that universal demonstrations are superior to particular ones. Philoponus' commentary on this chapter is an important source of his own metaphysical analysis of universal predications. After two chapters completing the discussion of the superiority of certain kinds of demonstrations (25-6), the remaining chapters of Book 1 of the *Posterior Analytics* offer a variety of topics, such as the precision of sciences (27), the scientific genus (28, 32), and the distinction between scientific understanding and other forms of cognition (among others perception, opinion, and acumen). Philoponus' commentary on these chapters is interesting for his views on the scientific genus and on intellect.

Philoponus and the Forms

Are there Forms such as those posited by Plato in some of his dialogues? If so, what metaphysical and epistemological role do they play? Is the existence of these Forms compatible with Aristotelian philosophy? If so, was Aristotle aware of this? These were all live questions among the Neoplatonic commentators on Aristotle. In different contexts Philoponus said different things. This invites a developmental account of Philoponus' thought concerning Forms, and an evolution of his understanding of the history of philosophy. One standard account of Philoponus' philosophical development[2] can be summarised as follows. In the beginning of his career, Philoponus followed his teacher Ammonius in seeing Plato and Aristotle as in agreement in regard to Forms. He later came to see them in disagreement. According to the later view of Philoponus, Aristotle does not believe that there are any Platonic Forms at all, not even as *logoi* (rational expressions) in the mind of God. Because God, on this new understanding, does not know the things in the physical world, He is not understood to be responsible for their existence. Hence, Philoponus comes to see Aristotle as positing a God that is a mere final, not efficient, cause of the world. On this developmentalist account, Philoponus comes to understand Aristotle's metaphysics in a way that brings it more in line with Christian teachings, according to which God is independent of the world, and can, and did, exist, apart from His creative activity. We here suggest that this account of Philoponus' development rests on a misreading of the evidence, including a crucial passage of the present commentary (242,14-243,25), which signals a new understanding of Aristotle's metaphysics as in opposition to Plato concerning the Forms. But this is because he now takes Aristotle to interpret Plato's Forms according to a literal reading of the *Timaeus*, according to which they are independent substances, prior to the Demiurgic Intellect. Philoponus *himself* continued to believe in Forms as 'demiurgic *logoi*' inherent within, and posterior to the Divine Intellect, and there is no clear evidence that he did not follow Ammonius in attributing this same view to Aristotle.

It is generally agreed that at least part of Philoponus' commentary on the *Posterior Analytics* dates from the earlier part of his career, as it is largely a record of Ammonius' lectures on the *Posterior Analytics*. But as we shall see, there is no question that Philoponus' understanding of what Aristotle had to say of Plato's Forms underwent some change from very early in Philoponus' career; what is at issue is exactly what that change is. It would be in order to briefly review the evidence. At *in DA* 37,16-38,16 (an uncontroversially early work[3]) Philoponus comments on Aristotle's assertion at *DA* 1.1, 402b7 that 'The living being in general is nothing or something posterior.' Philoponus points out that these lines could be, and have been, taken as evidence that Aristotle rejects Forms as Plato conceives them, but he here rejects this reading. He instead follows the entrenched tradition within the Neoplatonic study of Aristotle's treatises that took Plato and Aristotle to be in accord in regard to the main theses of their philosophies.[4] Philoponus defends those who take the approach of reconciling Aristotle with Platonism, as he, following Ammonius (who in turn follows Proclus) understands it: the Forms exist at two levels, as subsistent eternal beings prior to the Demiurge, to which the Demiurge's intellectual activity is directed, and as *logoi* within the Demiurgic Intellect, by which creation is accomplished. He offers the following considerations in support of this reconciliation. (1) Following Ammonius, Philoponus points to Aristotle *Metaph.* 12.10, 1075a11-25, in which Aristotle asserts that the universe contains the good both as something separate and as something internal, and, to show how this might be so, considers an army, of which the goodness consists both in the leader outside of it and the order within it. Aristotle elaborates on what he means by the order within it: there is an interconnectedness of all things in the cosmos, according to which lower beings are for the sake of higher beings, and so forth, all contributing to one end. Philoponus adopts a Platonic perspective according to which the commander is the Demiurge of the *Timaeus*; He has within his mind the Form of Animal itself, and the cosmos has its internal order by virtue of being created according to the exemplar of the Animal Itself. (2) Philoponus points to Aristotelian assertions that the intellect is identical with its objects, and that God is an intellect, for which reason God too is identical with His object of thought. 'Again, in the *Metaphysics*, when discussing the Divine Intellect, he says that the forms of all things are present in it; at any rate, he says that when seeing itself it sees all things, and when seeing all things it sees itself' (37,27-9). Philoponus' paraphrase employs a plural not present in the text of the *Metaphysics* itself, showing that he follows the Neoplatonic reinterpretation of the Divine Intellect as a unity embracing a plurality of intelligibles. Philoponus accordingly interprets Aristotle's 'The living being in general is nothing or something posterior' by appealing to the familiar Neoplatonic distinction between the Form as Divine exemplar and the 'deflated' Form as shared characteristic, whose universal characteristic is a consequence of the mind's consideration of particulars.[5] 'In general' here renders *to katholou*, which can also have the sense of 'universal'; for Philoponus, Aristotle is saying that there are no universals that exist in themselves, only shared characteristics, which are dependent on and necessarily inherent in particulars. (The same account of universals is attributed to Aristotle at 272,31-273,75. There, in a telling passage that reveals his lingering Platonic tendencies, Philoponus insists that, although shared characteristics are not subsistent, they are in a sense more real than the particulars in which they inhere, since they are eternal.) Nonetheless, according to Philoponus, those particulars themselves are what they are on account of intelligible objects which are themselves particulars of a sort. Philoponus takes it to be a misunderstanding to take Aristotle's arguments against the Forms as directed against the Forms as predemiurgic subsistent exemplars or as *logoi* in the mind of God, both of which, on his understanding here, Aristotle takes to exist. Philoponus here repeats and endorses what his teacher Ammonius had insisted: that Aristotle's arguments against the Forms are directed against those who take Forms to be universals subsisting apart from the level of Intellect. Plato himself did not understand the Forms in this way; hence Aristotle sees himself as presenting arguments against those who would distort Plato's teachings, not against those teachings themselves.

Against Proclus On the Eternity of the World, a late work, stands in contrast. Proclus had argued that Forms, which subsist eternally in the Divine Intellect, are essentially exemplars of particular things in the world. To be an exemplar is to stand in relation to that for which it stands as pattern. So, he argued, they must stand in an eternal relation with particulars in the world, for which reason the physical world, too, must be eternal. Philoponus begins his refutation of this argument by remarking that not everyone has agreed with Plato's account of the Forms and their ontological role, as is made apparent by Aristotle's frequent expressions of disagreement on the matter.[6] Philoponus cites *Metaph.* 1.6, 987a29-b10, where Aristotle identifies Plato as one who posited Forms as separate from perceptibles, as evidence that when Aristotle attacks the thesis that Forms are independent substances, he is taking issue not with the thought of those who misunderstood Plato, but with the views of Plato himself. Philoponus asserts that Plato clearly understood Forms to be substances, not *logoi* within the Divine Mind, and it is this authentically Platonic position – that of the existence of predemiurgic Forms – with which Aristotle is taking issue. Philoponus does not here consider how to deal with the texts he had previously adduced as evidence that Aristotle's beef was only with a misunderstanding of Plato, the view that there are separately existing universals, and not with postulating eternal patterns in the divine mind. But recall that Ammonius, following Proclus, posited Forms as both prior to the Demiurgic Mind and as inherent within it. The texts that Philoponus had adduced in *in DA* in support of the harmonisation of Plato and Aristotle purportedly showed that Aristotle believed in Forms as *logoi* within the Demiurgic Intellect. That is consistent with the rejection of Forms as separate substances. As before, Philoponus rejects the separate independent existence of Forms. But now Plato is interpreted as holding that there are indeed such subsistent

Forms. On the basis of these two texts, we can conclude that Philoponus changed his mind on what he took Aristotle to think Plato said about the Forms, but not to change his mind concerning Aristotle's, or his, view of the status of the Forms themselves.

The *Posterior Analytics* commentary discusses Plato's Forms in a way consistent with this later understanding. Philoponus comments on an Aristotelian passage that is among those that Philoponus cited in *Against Proclus* to support his contention that Aristotle did not believe in Forms: 'So goodbye to the Forms. For they are but noodling, and even if there were such things, they would be irrelevant in regard to the argument. For demonstrations are concerned with these sorts of things' (83a32-5). Philoponus remarks that, while Plato would take the Forms of nonsubstances to be themselves independent substances (and hence themselves such as can serve as the ultimate bottom of predication), Aristotle denies that there even are such things. He first lays out the case of those who would minimise the disagreement between Plato and Aristotle, citing some of the same texts that Philoponus himself pointed to in the commentary on *De Anima*, when he himself tried to harmonise Plato and Aristotle on the Forms. He concludes 'But this is not what Aristotle is up to, here. Rather, it is obvious that he is always doing battle against the doctrine [of Forms], not those who conceive of it incorrectly. For in the *Metaphysics* he draws out many long refutations of the doctrine ... And it is obvious that in reality Plato did not say that Forms are simply *logoi* within the demiurge, but he granted them subsistence in themselves and [said] that the equal itself and the animal itself and such things were something, and also said that it is in regard to these, as paradigms of images, that the Demiurge fabricated the things here' (243,17-25).

As in *Against Proclus,* Philoponus says Aristotle was absolutely right in taking Plato to be advocating the substantial existence of Forms and he was right in rejecting this theory. The conclusion follows that Aristotle is right in insisting that no nonsubstantial term can serve as minor term of a demonstration, if the predications that make up a demonstration are indeed 'natural'.

Philoponus does not specify which arguments in the *Metaphysics* he has in mind as evidence that Aristotle was unequivocally opposed to positing Platonic Forms. Presumably he has in mind the puzzles presented in 1.9 concerning the paradoxes of participation and self-predication, which arise from positing a Form as an independent entity, of the same kind as the things that have their character through their participation in that Form. Neither these puzzles nor any other in Aristotle of which we are aware would affect those who took Forms to be only ideas in the Divine Mind.[7]

Verrycken takes the passage at hand to be evidence that the *Posterior Analytics* commentary was revised later in Philoponus' career, and, since other texts within the same commentary suggest that the world is eternal, a key doctrine of the early Philoponus, he concludes that the commentary is a product of both earlier and later strata of Philoponus' thought.[8] But note what the passages from *in An. Post.* and *Against Proclus* do and do not state. Both passages affirm that Plato thought that Forms exist independently as subsistent in themselves, that Aristotle recognised that this is how Plato understood them, and that Aristotle explicitly departed from Plato on this point. Both passages affirm that Aristotle was right on both counts. Left unchallenged is the view that Forms subsist as *logoi* within the Divine Intellect, by virtue of which they serve as patterns by which particulars come to be.

There is evidence that Philoponus himself continued to believe in such Forms, even late in his philosophical career. In *Against Proclus On the Eternity of the World* Philoponus argues that positing a temporal beginning of the cosmos does not entail attributing potentiality to God. He writes 'And so, as has been shown, the creative principles (*logoi*) for things which are within God always possess actuality and perfection but God brings each thing into existence and gives it being when he so wishes.'[9] Philoponus' views on the Forms change only insofar as he no longer takes these *logoi* to be Forms as Plato understood them.

The following story therefore emerges. At one point Philoponus took Plato to speak of Forms as self-subsistent only as a manner of speaking; their true status was as inherent within the Divine Intellect. Plato was right in holding this view. Aristotle rightly interpreted Plato in this way, and agreed with Plato. At another point in his career Philoponus continues to take the paradigms by which things are what they are to be dependent on the Divine Mind in which they somehow subsist, but interprets Plato as giving to Forms an independent substantial subsistence, as they have on a literal reading of the *Timaeus*. Aristotle disagrees with Plato, on this. Unclear is whether he takes Aristotle to deny intelligible Forms as *logoi* within the Divine Intellect. The shift here concerns the interpretation of Plato and Aristotle; it is not a profound philosophical shift, and need not be correlated with more profound philosophical shifts later in Philoponus' career, such as that to the view that the cosmos has a temporal beginning.

Philoponus on the scientific genus

The genus of a science guarantees both its unity and its distinctness from other sciences. Nonetheless, in both Aristotle and later thinkers, many of the details concerning the nature of the genera of sciences are still unclear. In the following, we will briefly address Philoponus' notion of scientific genus, starting from the following three questions: What guarantees the unity of a science? What distinguishes sciences from one another? And what is a scientific genus? In the process, we will address the role of the scientific genus in the common notions. Although Philoponus, like Aristotle, does not present a definite and clear answer to the questions mentioned, we can nonetheless say that his explanation is somewhat more elaborate and informative than that of Aristotle.

With respect to the unity of a science, Philoponus takes a more extreme position than Aristotle. Early on in his comments, in discussing the subalternate sciences (75b15 with 100,7-32), he adopts what we may call a standard view by taking over Aristotle's description of them as having the same genus 'in a sense' (101,19-20). Later on in the commentary, he further develops this view when he states that the relation between a subordinate/more specific science and a superordinate/more general science is one of inclusion and causal relation: the superordinate science includes the subordinate one, of which it is the cause.[10] When explaining 1.28, however, Philoponus takes a step further, and goes beyond Aristotle by understanding his statement 'a science is one if it has one genus' quite literally. Philoponus concludes that in that case the subalternate sciences are in fact one and the same science: they concern one genus, including its subspecies, and hence are one science. Sciences are different, he says, only if they do not use the same (proper) principles and the one does not use the theorems of the other as principles, and therefore geometry, stereometry and optics, insofar as they all study 'the continuous, or magnitude, and the species thereof', are the same science (302,12-15, cf. 303,12). The fact that stereometry and optics study *species* of the genus does not make them separate sciences.

So what is a scientific genus according to Philoponus, and how does it guarantee the distinctness of sciences? Unfortunately, like Aristotle, Philoponus is not very clear on this issue. What is clear, however, is that he assumes an important role is set aside for the categories. At 313,30ff. (ad 88a30ff.), Philoponus distinguishes the scientific genus of musical theorems from that of geometrical and arithmetical ones, by pointing out that the former belong to quality, rather than quantity. He then goes on to state that the problems of medicine deal with yet another genus, namely substance (or more precisely, substance plus matter, see 314,9). This identification of quality, quantity, and substance as scientific genera is clearly inspired by the Aristotelian categories. However, the identification of genera with categories cannot be understood as implying a definition of scientific genus as category. Earlier in his work, when commenting upon Aristotle's formulation of the prohibition of kind-crossing at 84b14ff., Philoponus explicitly states that 'of the same genus' does not simply mean 'falling under the same category', as one category can be the subject of more than one science. Philoponus there rather uninformatively concludes that the term genus must be understood as referring to 'that which belongs to a science', or 'the classes (*merê*) found' in a science (265,22-5). The examples he mentions, 'lines or shapes or some such thing', suggest that he is thinking of lower species falling under a category, in this case, continuous quantities.

More informative with regard to the relation between the categories and the scientific genus is Philoponus' analysis of what distinguishes geometry and arithmetic, which are probably the clearest example of two sciences studying one category, namely quantity. To be more precise, these two sciences both study a species of quantity, but this does make them separate sciences, because the species in question are the first, and mutually opposed, species into which a genus (in the sense of category) is divided (314,4-5).[11] As a consequence, it seems Philoponus would not be able to agree with Aristotle that quantity is a univocal genus, but instead agrees with Zabarella that it is equivocal.[12] Moreover, in that case Philoponus cannot follow Aristotle in assuming a universal science of quantity.

Although this idea of opposite species functioning as genera of distinct sciences is not elaborated further, there is one other remark in the part of the commentary here translated which suggests that the idea is crucial for Philoponus: he understands 88b20-1 to mean that there is one primary immediate premise for every genus of a science, and concludes from this that the ten categories cannot be equated with the scientific genera, as every category falls into two opposed kinds, each with its own primary immediate premise (320,11-15). This seems to imply that there are two distinct, non-subordinate sciences for each category, each with one corresponding primary immediate premise.

Philoponus aims for a comprehensive account of those elements of a science in which the genus plays a role. His discussion ranges from the very first principles of a science or the common notions [13] through the so-called proximate principles[14] (i.e. theorems and terms)[15] all the way to the individuals, i.e. the individual objects a science is about (265,26-266,2). We will here discuss only Philoponus' view of the top of the chain, the first principles or the 'universal and common notions'.[16] On this level, the scientific genus is relevant in two senses. First of all, when Aristotle says that 'not even ... among the common principles there are some from which everything can be proved', since 'things are proved through the common principles with these [things[17]]' (i.e. things belonging to a specific genus), he means, according to Philoponus, that we always need a premise 'from the genus that underlies' the science in question, next to the common notions. On the basis of common notions alone one cannot prove anything, since they are not about any particular kind of thing.[18] Second, as a consequence, the common notions themselves need to be made to be about something, or more specifically need to be adjusted to the science in question.[19] Thus, Philoponus says, the manner in which the sciences employ the common notions is not common to all sciences, but each science adjusts them to make them appropriate to its genus.[20] It does so by inserting the relevant subject into the common notion: in the example 'whenever any four things are proportional, they will also be alternately proportional', the neutral 'things' (*tina*) becomes e.g. 'magnitudes', or 'numbers', or 'intervals of time'.[21] So when functioning within a science, the common notions are literally only about the subject matter of the science in question. Taken in themselves, however, they are not about any subject matter.

This relates to a vexing issue in Aristotle's *Posterior Analytics*, namely whether the commonness of the common notions extends to all sciences or only to some. Philoponus gives different answers to this question in different phases of the commentary, but his general position towards the end of Book 1 differs from Aristotle's in that he takes the commonness to extend to all sciences. Early on in his commentary on Book 1, Philoponus distinguishes very clearly between common notions which are relevant to all, some, or even only one science (10,27-11,3).[22] Further down the road, however, at 315,25-316,22, where he wants to contrast the universality of the common principles with the need for a premise from the relevant scientific genus, he mentions only common notions which are used by all sciences, namely the Law of the Excluded Middle and that of the transitivity of equality. Likewise, at 321,11 Philoponus maintains that the common principles are 'the same' in the sense that they are used by all sciences (cf. 266,22, where he explains 'most universal immediate premises' as referring to the common notions).[23] This position regarding the common notions, which is a bit more outspoken than that of Aristotle, fits the view described above that, taken in themselves, the common notions are not about any subject matter (otherwise how could they be used by all sciences?). Moreover, it fits Philoponus' view of the role of metaphysics and intellect (*nous*) as providing the principles of all sciences.

Philoponus on intellect

First philosophy or metaphysics is the science which 'is common to all sciences', as 'the art of arts and the science of sciences' which 'discovers and demonstrates the principles of every science'.[24] A very similar, if not the same task, is set aside for intellect.[25] The section of the commentary translated in this volume is interesting in this context because in it we find Philoponus' answer to the question whether the activities of intellect involve propositional thought.[26]

Intellect is the capacity of the soul with which we 'deduce' the universal from the perceptions or particulars (308-9 n. 409). Aristotle's statement that intellect is that by which we know the *horoi* (terms or definitions),[27] is explained by Philoponus as referring to our knowledge of 'the divine and intelligible forms' (324,6-9).[28] Moreover, intellect is that which discovers the categories (332 n. 537) and induces their number (239,9). These are all unitary and simple entities, fitting the unitary manner of knowing of intellect: intellect apprehends the forms through so-called simple intuitions (*haplai epibolai*, 324,6).[29]

That does not imply, however, that intellect does not know propositions. In fact, Philoponus locates the apprehension of immediate premises and common notions – both of which are propositional – in intellect as well. The question arises, then, if knowing propositions is compatible with intellect's unitary way of knowing. Or to put it differently: does intellect know the immediate premises and common notions in a unitary fashion?[30] The knowledge intellect has of the common notions and of the immediate premises is superior to demonstrative knowledge in that it grasps them immediately, rather than on the basis of prior knowledge.[31] But that does not necessarily imply that it knows them unitarily, i.e. non-discursively.

For an answer to these questions, the commentary on 2.19 is hardly helpful. In it, Philoponus[32] emphasises the superiority of intellect over scientific understanding (*epistêmê*), which is due to the fact that the former provides the principles of the latter and thereby is itself a kind of principle of scientific understanding. The principles in question, according to Philoponus' exegesis of the chapter, are the definitions (438,22-3[33]) and the immediate premises (e.g. 440,9). We are not informed about whether they are known unitarily or not. In Philoponus' *in De Anima*, we read that intellect thinks that which has parts as having parts, but not as composite, whereas *dianoia* thinks it as composite (esp. 545.33ff.).[34] So definitions, which are not propositions, may not be a problem. The question remains, then, in what way common notions and immediate premises are known by intellect.

The commentary on *An. Post.* 1.3 clearly speaks of a propositional understanding of intellect. Philoponus there follows Themistius in maintaining that intellect, as the principle of scientific understanding, knows the terms (*horoi*) of which the common notions (here called *axiômata*) are composed. Interestingly, however, intellect not only knows these terms, by simple intuitions (*haplai epibolai*), but also combines them (*sumplekei*) into common notions (*in An. Post.* 48,7-17), such as the Law of the Excluded Middle. That the resulting combination is considered to be propositional, is clear from the description of the elements thereof as 'subject' and 'predicate' (48,10).

How is it possible for one capacity of the soul, intellect, to have both a unitary, non-propositional or non-discursive, grasp of terms, and a propositional grasp of common notions? This becomes clear in Philoponus' commentary on *An. Post.* 1.33. Philoponus presents a rather Neoplatonic solution which allows intellect to also think discursively, by distinguishing different layers in the activity of intellect. As it turns out, it is not intellect as a whole which apprehends the divine and intelligible forms, but only what Philoponus calls its summit or perfection, i.e. wisdom (*sophia,* 332,6-8), which comes to souls which have been perfected (332,21-2). To this task of *sophia* Philoponus opposes the task of intellect, i.e. the rest of intellect, which comes down to apprehending the immediate premises, and discovering the ten categories (332,25ff.).[35] Most interestingly, Philoponus also distinguishes a 'last and lowly activity' of intellect, by which it knows immediate premises and common notions (325, 9-12). For Philoponus, then, it is no problem that one cognitive capacity knows in different ways, since he distinguishes layers of activity within that one capacity.[36]

Acknowledgements

We thank Richard Sorabji for inviting us to translate this text, Michael Griffin, Sebastian Gertz, and Katharine O'Reilly for their help in preparing our translation for publication, and Mariska Leunissen, Richard McKirahan, Orna Harari, and Jonathan Barnes for their valuable comments. Owen Goldin would like to thank his research assistant, Damon Watson for his sharp eye and sound judgment, and his wife Miriam Sushman, for good things without *peras*. Marije Martijn would like to thank the research group Ancient and Medieval Philosophy VU University Amsterdam and Wouter Goris and Arianna Betti for their helpful questions and suggestions. Marije Martijn's work on this volume was made possible by a grant from NWO, Veni-project 275-20-020.

Marije Martijn dedicates this volume to her parents; Owen Goldin, to his wife Miriam.

Notes

1. See Richard (1950).

2. Verrycken (1990b), (1994), (1998).

3. Verrycken (1990b), 255-7; van der Eijk (2005), 2.

4. See *in DA* 37,18ff. Sorabji (1990), 5 aptly calls this 'a perfectly crazy position' that 'proved philosophically fruitful' as philosophers worked through the implications and consequences of such a reconciliation.

5. On this distinction see Sorabji (2005), vol. 3, 128-63; Sorabji (2006).

7. At *Met.* 1.9, 991a20-3 Aristotle does take issue with calling the forms paradigms, on the basis of how poetical language leaves mysterious the ontological relations in question.

8. Verrycken (1990b), 257.

9. *In Aet.* 79, 4-7, tr. from Share (2005), 64.

10. 117,15 (ad 1.9). This may betray a Neoplatonic view of genus as including all the species. The higher science is itself called the cause of the lower science (rather than providing the causes, vs. providing the facts) at 301,5.

11. These species are in turn called their genus, in the sense of scientific genus, 304,8.

12. Cf. McKirahan (1992), 72 and see below, on common notions.

13. Common notions: 315,26-316,21.

14. On the distinction between first and proximate principles see 311,26-312,2; 313,29-30.

15. 88a30ff. Philoponus on proximate principles as theorems: 314,4-12.17-19; as the terms of which the theorems consist: 314,14-16.

16. Philoponus calls them 'common notions', 'common axioms', and simply 'axioms'. We will use the expression 'common notions' (*koinai ennoiai*), which is used most in the part of the commentary translated in this volume.

17. On this neutral translation see n. 455 to the translation.

18. 315,27-316,14. Considering the formulation Philoponus uses for this argument, this reading must be inspired by 88b27-9, where Aristotle distinguishes between the principles 'from which', which are common (*koinai*), and the principles about which, which are specific (*idiai*). See esp. 316,4-5 and 321,8-9.

19. This is Aristotle's own position. The issue first arises in relation to *An. Post.* 1.10 76a37-b2, although Philoponus brings it in right in the introduction to his commentary, as he takes the common notions to fall under pre-existing knowledge (3,23-4). See McKirahan (2008), 17.

20. 316,14-22.

21. This position may be based on Philoponus' reading of Aristotle's statement that the common notions are applicable in different sciences *kat'analogian* (*An. Post.* 1.10). Philoponus takes this to mean that the common notions are the same for all sciences, but only homonymously. See McKirahan (forthcoming) and cf. McKirahan (1992), 72 on Zabarella's interpretation of common principles as equivocal.

22. See McKirahan (2008), 23 and notes, and, for a critical note on the view that some axioms are relevant to one science, McKirahan (1992), 69 and n. 7. Note that Philoponus also has a different explanation of the commonness, namely that the common notions are shared by everyone. On this topic see McKirahan (2009) and Martijn (2010), 112ff.

23. At 320,28 he seems to maintain that some common notions are useful in some sciences, and others in others, but this need not be Philoponus' own position, as the context is hypothetical ('maybe someone will say').

24. 118, ad *An. Post.* 1.9, 76a15, 'the principles of [the subalternate sciences] have a common feature.' This common feature, according to Philoponus, is first philosophy.

25. The relation between first philosophy and intellect will not be discussed here.

26. On this question in Aristotle, Plotinus and Proclus, see Sorabji (1982) and Sorabji (2005), vol. 1, 91ff.

27. *An. Post.* 1.3, 72b4 and 2.19, 100b15.

28. This is probably why Philoponus emphasises that intellect is discussed in Aristotle's theological works, namely the *Metaphysics*. See 331,10-12 and note.

29. 'Simple intuitions' (*haplai epibolai*) is a Neoplatonic technical term with roots in Epicurean and Platonic epistemology. See further n. 500 to the translation.

30. For a recent discussion of the question whether Aristotle, in *An. Post.*

2.19, wants *nous* to know both concepts and propositions, see Perelmuter (2010).

31. 254,29-255,2. Cf. 308,6ff.: in general, intellection (*noêsis* used in a broad sense) with an explanation is superior to intellection without an explanation. However, intellection of the common notions, as well as intellection of the first cause, which are first principles (in different senses) and hence cannot be known with an explanation, is superior to intellection with an explanation.

32. We will assume for now that the commentary at least derives from Ammonius' school.

33. Note that Philoponus here switches from universal concepts to the immediate premises and definitions as universals. The universal concepts in question are the Neoplatonic 'one over many', as opposed to the 'one before the many' and the 'one after the many'.

34. Philoponus here ascribes the thinking of propositions (*protaseis*) to thought (*dianoia*) and the thinking of terms (*horoi*) to intellect, but this may be no more than exegesis of the relevant passage from *DA*. For another option, see n. 32 to this introduction. For the simplicity of definitions, cf. *in An. Post.* 2, 339,6-11.

35. Actually, Philoponus in this passage also ascribes apprehension of the terms or definitions to intellect as opposed to wisdom, referring again to the above mentioned passage from the *An. Post.* Earlier, however, those terms or definitions were explained as referring to the intelligible forms – which, as he just said, are apprehended by the summit of intellect, wisdom.

36. A similar, and possibly even the same, division of labour, is found in the *in De Anima*, between the actual and the potential intellect (the latter is equated with *dianoia*, *in DA* 491,5ff.).

Owen Goldin is chiefly responsible for the translation of the commentary on *An. Post.* 1.19-25, and Marije Martijn for 1.26-34. The present volume is however a collaboration, for which we take joint responsibility.

Textual Information

Textual emendations

This is a translation of the text printed in *Ioannis Philoponi in Aristotelis Analytica Posteriora Commentaria*, ed. W. Wallies, *CAG* 13.3 (Berlin: Reimer, 1909), with the following emendations.

218,16 Reading *hôste an eiê dialektikos* with R.
219,10 Repunctuating from a question mark to a full stop.
220,23-4 With the Aldine text, reading *proionta dia tôn mesôn*, moving it to after *katêgoroumena.*
226,4 Deleting the first *de* and reinstating the second, which Wallies deletes.
241,10 Reading *to katêgoroumenon ti* with the Aldine text.
243,24-5 Following Wallies' conjecture of *hôs paradeigmata eikonôn* for *eis paradeigma ex ekeinôn.*
252,5 Reading *tou leukou to khrôma* with R.
259,9 Reading *endekhesthai* for Wallies' *endekhethai.*
261.16 Reading *pleiosi* with G and a2.
265,17 Reading *tou hupokeimenou* for *to hupokeimenon.*
266,17 Reading *deiknutai* for *deiknuntai.*
288,26 Reading *tôn* for *tou.*
289,13 Reading C, C, B, B with RUA2 for B, B, C, C.
291,4-5 Transposing *ek pollou* with the Aldine text and reading *dêlon hoti ek pollou tou periontos* instead of *ek pollou dêlon hoti tou periontos.*
292,1 Not accepting Wallies' addition of *on.*
298,2 Accepting the *oude* omitted by Wallies.
299,18 Accepting Wallies' insertion of *to*, but rejecting his deletion of *sphairoeidês.*
303,5 Reading *haterai* or *hai heterai.*
305,2 Reading *ti* for *tí.*
316,10 Accepting the emendation proposed by Wallies and reading *ekeinai isai* for *monai isai.*
333,10 Adding commas before and after *phêsin.*

Notes on the text of Aristotle's *Posterior Analytics*

Discrepancies between the text of the lemmas of the *Posterior Analytics* quoted in Philoponus' commentary and the text as given in Ross' edition:

81b20-1 The lemma has *hôste ei kai esti têi alêtheiai tôn AB meson, dokei de mê,* while Ross has *hôste ei kai mê esti ti têi alêtheiai tôn AB meson, dokei de einai.*
82a8 Ross has *perainetai* for the lemma's *perainei.*
82a17-18 Ross reads *eit' … eit' …* for *ei t' … ep' …* of the lemma.
82b12-13 Ross has *hôste epei hê epi to anô istatai hodos, kai hê epi to A stêsetai* for the lemma's *hôst' epei hê epi to katô istatai hodos, kai hê epi to anô stêsetai.*
82b32 Ross has *peperasmenakis* for the lemma's *peperasmenôs.*
83b14 Ross reads *ti tôn toioutôn,* instead of the lemma's *tôn toioutôn.*
83b17 Ross has *dê* for the lemma's *de.*
83b24 Ross has *touto kath' heterou* for the lemma's *all' atta kath' heterou.*
83b33 Ross has *esti* for the lemma's *estai.*
84a3 Ross has *dielthein* for the lemma's *diexelthein.*
84a13 Ross has *ekeinois enuparkhei* for the lemma's *en ekeinois huparkhei.*
84a19 Ross has *an* for the lemma's *an en.*
84a20 Ross puts the comma after *esti.*
84a22 Ross reads *en tôi heni* for the lemma's *tôi heni.*
84b26 Ross reads *all' arkhê, kai* while the lemma has *all' arkhai kai.*
84b30 Ross has *hôsth'* for the lemma's *hôste.*
85a9 Ross excises *ê mê panti,* included in the lemma.
85a12 Ross has *badieitai* for the lemma's *peseitai.*
85a25 Ross reads *anthrôpos* for the lemma's *ho anthrôpos.*
85a33 Ross has *tautên* instead of *toiautên* of the lemma.
85b4 Ross punctuates as a question.
85b38 Ross has does not have the lemma's comma before *hoti allo.*
87a24 Ross reads *BC* for *AB* of the lemma
88a14 Ross has *hualon* for *huelon* of the lemma.
88a16 Ross has *têi horan* for the lemma's *dia to horan.*
88a35 Ross has *harmottein* for *epharmottein* of the lemma.
89a11 Ross omits *ouk,* found in the lemma.
89a22-3 Ross punctuates as a question.
89b15 Ross brackets *ho,* found in the lemma.

PHILOPONUS
On Aristotle Posterior Analytics 1.19-34

Translation

John of Alexandria's lecture notes from the meetings of Ammonius, son of Hermeias, on the first [book] of Aristotle's *Posterior Analytics*, together with some observations of his own

Chapter 19

81b10 Every syllogism[1] [comes about] through three terms, 216,27
[and one [syllogism] is able to prove that A belongs to C through [the fact] that it belongs to B and this to C, while another is privative, and has as one premise that something belongs to something else, and as the other premise that [something] does not belong.]

Here he intends to teach us a theorem[2] of great beauty, I mean, that demonstration cannot proceed to infinity, that is, that those who take up a given conclusion and link another term to it cannot produce a syllogism to infinity.[3] He proves this by proving that predications do 217,1
not proceed to infinity. For if they proceeded to infinity, it would be entirely necessary that demonstrations too proceed to infinity.[4] But since they do not proceed to infinity (rather, they stop and there are most generic genera of [predications]), clearly, demonstrations too cannot be infinite. The philosopher[5] said that the aim of the [lines] before us is to prove that there are immediate premises.[6] Since he[7] said that demonstrations must be based on immediate premises, he now proposes to prove this very [point] (that there are immediate premises). He proves this by proving that predications do not proceed to infinity. This is true for immediate negations. For there cannot be immediate negations unless there are certain genera that are the most universal. For it was asserted that those negations are immediate, which result when most universal genera are denied of each other.[8] Nonetheless, even if there are infinitely many predications, nothing prevents the existence of immediate affirmations.[9] For [when we investigate] immediate affirmations we are not investigating the most universal items as such, rather [we are investigating] those items that belong to others adjacently, with the result that a term cannot be inserted between them. For human being is immediately affirmed of Socrates and rational of human being, and animal of this. For an intermediate term to which the predicate primarily belongs cannot be inserted. But if one must speak more precisely, both of these [points], that demonstrations do not proceed to infinity

and that there are immediate premises, are proven through each other. For if there are immediate premises, demonstrations cannot proceed to infinity, and if demonstrations do not go on to infinity, it is entirely necessary that they end in immediate premises. Now this is the aim of the [text] before us. First he recalls what was previously said by him concerning second figure syllogisms. For he makes use of these [points] in the present [passage] too. For he wants to prove that it is not possible to predicate some things of other things to infinity.[10] Now, he says, a syllogism [comes about] through three terms, and either from [premises] of which either both are affirmative, or the major is negative. Now he will prove that a negative premise cannot proceed to infinity with the result that one could always find something denied of the subject and that this cannot [occur] for an affirmative [premise] either. For example, suppose that A belongs to no B, and B to C, and this to D, and so on to infinity (and likewise in the case of affirmatives).

218,1 > **81b14** So clearly the principles and the so-called hypotheses are the same. [For you must assume them and prove them as follows …]

He calls the premises 'principles' and 'hypotheses'. For there is a syllogism when these have been posited.[11]

> **81b16** … for example that A belongs to C through B [and again, that A belongs to B through another middle, and that B likewise belongs to C.]

By proving [not only] that A belongs to every B, but that B too belongs to every C, he proved that A also belongs to every C. But even if we wanted to prove each of the premises as well, we would do the same thing again by taking a middle term. Now he investigates whether it is impossible to do this to infinity.

> **81b18** In the case of those who deduce in regard to opinion, that is, in a merely dialectical manner, it is clear that this alone is to be sought: whether the [premises] on which the syllogism is based are the most reputable ones possible. So even if in truth there is a middle term for AB, but there does not seem to be one,[12] the one who deduces through this [deduction[13]] has deduced dialectically. [But for truth one must look into things on the basis of [premises] that hold.]

Since he said that demonstrations are not generated in any other manner than through immediate premises, he says that, even if some people deduce in a dialectical manner – since, as we saw, dialectical deductions are not from necessary [premises] but from reputable ones,[14] if the premises that are taken are not immediate, but seem immediate to the interlocutor, such a person nonetheless deduces in a dialectical manner. For dialectic is based on reputable deductions, which entails that [the deduction] is dialectical.[15] For example, [suppose that] someone were to speak as follows: 'soul is always in motion, what is always in motion is immortal, therefore soul is immortal'. Should it seem that 'the soul is always in motion' is [a] reputable [proposition] and that no middle term is required for its demonstration, that which seems [to be the case] will be taken as immediate. Likewise, even if, again, it should seem to be the case that the soul is immediately self-moving, that which seems to be the case will be taken as necessary. This is how the dialectical [deduction] is. However, the demonstrative [deduction] will take as immediate not that which seems immediate, but that which is immediate by nature.[16]

> **81b24** [This is how things are.] Since there is that which is itself predicated of another not accidentally, [– I mean this by 'accidentally:' when we say, for example, that that pale is a human being, we do not speak as we do when we say that the human being is pale …]

Since some predications are natural and some are unnatural[17] – and the natural [predications] are those which predicate the more universal of the more particular, or accidents of substances, and the unnatural are those which make the accident the subject and make the substance the predicate, as when we say that the pale is a human being, that the black is a crow (and such are unnatural because substances, in which accidents subsist, by nature serve as subjects[18] – and when we make the accident, which does not by nature serve as subject, [into] the subject, and we predicate of this the substance, which does by nature serve as subject, we produce an unnatural predication), because he is investigating whether predications proceed to infinity, it is to be expected that he first determines that his account concerns whether natural (as opposed to unnatural) predications proceed to infinity. (This is so whether we predicate accidents of substances or the more universal ones of the more particular, for example, rational of human being, and animal of this, and living of this, and so on to infinity, or it stops.)[19] Similarly, if we say that Socrates is pale and we predicate colour of it and quality of that, does this too [go] to infinity or does it too stop? It is to be expected that he produces his account in regard to such predications, too, since in the sciences we deal with the per se accidents[20] of things, too.[21] For example, if every triangle has three angles equal to two right angles, and these two are equal to adjacent angles, does this too [proceed] to

infinity or does it stop somewhere? So the aim of the text before us is clear. Within it he made use of three hypotheses. For he begins by investigating whether, when there is a determinate last subject, for example, the most specific or the individual species, the predicates
20 can proceed to infinity. For example, [is this possible] if Socrates is the last subject, and human being is predicated of him and rational of it, and then animal, and living, and so to infinity?[22] Second, <he investigates> whether, if there is a determinate last predicate, for example, a most generic genus, such that nothing else is predicated
25 of it, the way downwards proceeds to infinity.[23] Third, [he investigates] whether, when there are determinate extremes (I refer to both subject and predicate), the middle terms are infinite, <and> [whether], just as in the case of, say, continua, when the limits are determinate the items between [them] are infinite in number by division, so too, in the case of predications, when there is both a determinate last subject (for example, human being) and a last
30 predicate, for example, substance, the items between them are infinite in number.[24] So should one want to ascend through the continuum from the last item to the first, he would not be able to [do so], on account of the infinite [number of items]. So if it has been proven that the predications are infinite neither upwards nor downwards nor in the middle, it is pre-eminently clear that there is no way in which predications can proceed to infinity. Since this is so, demonstrations
35 too do not proceed to infinity. And this is why in the introductory chapters,[25] when responding to those who said that everything is
220,1 demonstrable, he said that for those who speak in this way the consequence will be not that everything is demonstrable, but that there will be a demonstration of nothing at all, if demonstration really is based on principles and (because the predications proceed to infinity) it is not possible to grasp the principles.

81b27 For he is not pale, being something else, but the pale [is
5 a human being] because it is an accident that being pale is in human being. [Now there are some things that are such as to be predicated per se.]

When we say that a human being is pale, a human being is thereby said to be pale, while he is not something else, for a human being is self-subsistent. So if a human being is self-subsistent, and the pale belongs to him, it is in a natural manner that we are positing human
10 being and are predicating pale of him. But when we say that the pale is a human being, since something else must be prior to the pale, and it subsequently becomes pale – yet we posit the pale as subsistent in itself[26] and subsequently predicate of it precisely that which by nature is its subject – it is reasonable to say that we produce an unnatural predication.

> **81b30** Let C be the sort of thing which itself no longer belongs to anything else; B belongs to it primarily, and there is nothing in between. [And again E belongs to F in the same manner, and this to B. So must this therefore stop, or can it go to infinity?]

The first hypothesis [is hypothesised] in order to see whether or not the predicates while proceeding through immediates[27] are infinite in number, should there be a determinate last item, which only serves as subject and is not in any way predicated of anything. For example, suppose that C is the subject, so that we take what is immediately predicated of it and again what is immediately predicated of that, etc.

> **81b33** And again, if nothing is predicated per se of A [but A is the first that belongs to H, and there is nothing prior in between, and H [is the first to belong] to G, and this to B, does it follow that this too must stop, or can this too proceed to infinity? This differs from the previous case insofar as the one is: is it possible for the one who starts from the sort of thing that does not belong to anything else but is such that something else belongs to it, to proceed upwards to infinity, and the other is to see whether it is possible for the one who begins from the sort of thing that is itself [predicated of] another, but has nothing predicated of it, to proceed downward to infinity.]

The second hypothesis hypothesises a determinate predicate which is such that it is not possible to predicate anything of it, and investigates whether as it proceeds through the immediates, it proceeds to the subjects, to infinity.

> **82a2** Further, can the items in between be infinite when the extremes are determinate? [I mean, for example, if A belongs to C and B is their middle, and there are other [items between] B and A, and others [between] these, can these proceed to infinity, or is this impossible?]

The third hypothesis [hypothesises] that the most generic item is determinate, just as it is the most specific, and investigates whether 221,1
or not the intermediates are infinite in number, with the result that neither would the predicate that first proceeds[28] through middle terms ever [reach] the most specific [one], which we saw to be only a subject, nor would the most specific [one] that first ascends through the middles ever reach the most generic item. Were this hypothesis true there could be no immediate premise. Yet it is not possible to insert an intermediate term for any given [premise], since, if the limits are determinate and it is possible for something to serve as an immediate subject for what is ultimately first (I mean for the most

generic item), and it is possible to predicate, a finite number of times, each of the things that have been taken of something, it is entirely necessary that the middle terms be finite in number. For if it is possible to immediately predicate something of everything that is taken, surely it is clear that there is a finite number of immediate predications, for which reason the middle terms are not infinite in number. But if they were infinite, it would not be possible to grasp an immediate predication. Now, we will consider later[29] whether or not this is true, I mean whether, when the extremes are determinate and the intermediates are infinite in number, it is not possible to immediately predicate one thing of another. When he says that the intermediates are infinite in number he means that whenever a premise is taken it is possible to take a middle term and that it is never possible to take an immediate premise.

> **82a6** Looking into this is the same as looking into whether demonstrations go to infinity, and whether there is a demonstration of everything, or whether [the extremes][30] are bounded[31] by one another.

With these [words] he clarified for us the aim of the [text] before us. If we make predications proceed to infinity,[32] demonstrations too must proceed to infinity, and when we are adding a term to the conclusion, there will always be a syllogism out of such [terms],[33] and if demonstrations go on to infinity, predications too must proceed to infinity, and it belongs to one and the same inquiry to investigate each of them. They 'are bounded by one another': that is, the extremes that we want to demonstrate are bounded by each other, and the predicate that first proceeds through the middle terms can [proceed] to the subject. Likewise, the subject that first ascends through the middle terms can [proceed to] the predicate, for if the middle terms are bounded, it is obvious that the middle terms are not infinite. He says that this[34] is the same as investigating whether there is a demonstration of everything. For if in the case of any given thing, it is possible to find something more universal (and this is what would happen, if the predications go on to infinity), and demonstrations are based on the most universal [predications], clearly, it is not possible to demonstrate everything, but the truth is to say that [it is possible to demonstrate] not even one thing, if demonstrations in the strict sense are indeed based on the most primary principles, and it is not possible to grasp the primary principles on account of how the predications proceed to infinity.[35] For if what is taken for the demonstration of something else needs to be demonstrated, and again what is taken for the demonstration of that needs to be demonstrated, and so on to infinity, and [if] it is not possible to assume the principle (if the principle is not only not proven, but is not such as can be proven), then, it is obvious that there is nothing of which there is a demonstration.

> **82a9** I say that this is how things are for both the privative syllogisms and the premises, [for example, if A belongs to no B, either it is the first to which it does not [belong], or there will be a prior intermediate to which it does not belong …]

He says that just as we investigate whether it is possible to predicate one thing of another to infinity, so [we can investigate] in regard to negations whether it is possible for negations to proceed to infinity. An example [of a negation] would be 'A belongs to no B'. If A does not belong to B, primarily, clearly negation does not proceed to infinity. But if it is not possible to deny something of something else primarily, negations must go to infinity. For example let A not belong to B, through the middle C. So if A belongs to none of the Cs, and C to every B, A belongs to none of the Bs. Again, let A belong to none of the Cs, through the middle D. If A belongs to none of the Ds, and D to every C, A belongs to none of the Cs. Let it be likewise proven that A does not belong to D through the middle E. Clearly this again [proceeds] to infinity. So when there is no immediate negation the affirmations cannot be finite in number, but they can proceed to infinity. But if there is a finite number of affirmations the number of negations too must surely be finite. And there is an immediate negation since it is not possible for a negative premise to be proven without an affirmative one.[36]

> **82a11** For example, suppose [that it is] G, which belongs to every B [and yet again to something else prior to this, for example to H, which belongs to every G. For in these cases too either those prior items to which it belongs are infinite in number, or they stop.]

He says that it must be investigated whether, if A is denied of B not primarily but through some other middle, for example G, this very G must no doubt belong to all B. For in this way, if A belongs to no G, and G to every B, A is denied of every B through the middle G.

> **82a15** This is not how things are for the items that convert. For 223,1
> among the mutually predicated items there is nothing of which something is predicated primarily or terminally.

He says that in the case of items that convert it is not possible to investigate the same things that we investigated in the case of items that do not convert. For among the items that convert neither a last subject nor a last predicate are determined. For when everything converts with everything else, if you take the last subject, you have thereupon taken the last predicate (for when predications convert, the last subject becomes the last predicate). For the same reason, if you take the last predicate, you thereupon also have the last subject. For example, capable of laughter and receptive of intellect and science, walking upright, human being, flatnailed, and mortal among rational beings convert with one another. Now among these it is not possible to take either a last subject or a last predicate, for if you were to posit for yourself a last subject, that could be a last predicate too. So if there is an infinity upwards, there must surely be one downwards too, and if downwards, there must surely be one upwards as well. For example, if human being is the last subject, then capable of laughter would belong to every [instance of] this, and receptive of intellect and science to this, and rational mortal to this, and walking upright to this, and flatnailed and all of the other items that you might think of that are coextensive with them, to this. But again, if you posit flatnailed as last you will find human being to be a predicate, and each of the rest of them will have the same status in regard to all of the rest. So concerning these things, there is only one inquiry, whether it is possible to find an infinite number [of items] that are coextensive with one another.

> **82a17** For in this way all of them stand in the same in relation to all.

That is, by [all] being taken as subjects or as predicates, all have the same status in regard to all.

> **82a18** And if it has infinitely many predicates, the things about which we are puzzling are infinite in both directions …

That is, if, when something has been taken, its predicates are infinite, [the items] about which we are puzzling must exist in both directions [since], if it is possible to predicate an infinite number [of
224,1 things], it is also possible for those who descend downwards to take certain subjects to infinity.[37]

> **82a19** … unless it is possible for them to not convert in the same manner, but the one does so as an accident and the other as a predicate.[38]

He says that in the case of coextensive terms, the very thing that is an accident of one or another of the extremes, this, as we mentioned, is surely [an accident] of the other, on account of the fact that any coextensive item is similarly disposed towards all. And they differ from one another only insofar as, among the coextensive items some are substances and some are per se accidents of substances.[39] And in this respect at least the conversion of predications is not similar, but when an accident (such as capable of laughter, say, or walking upright) is predicated of a substance, like human being, which serves as a subject, we produce a natural predication. But when capable of laughter or walking upright is the subject, and human being is predicated, it is unnatural. He called an unnatural predicate an accident, and he accordingly [called] a natural [predicate] a predicate, without qualification. For he intends to soon make a division of predicates,[40] and all of those that are predicated of something naturally, he calls predicates, and all of those that [are predicated] unnaturally, he calls accidental predicates.

Chapter 20

> **82a21** So it is clear that the intermediates cannot be infinite in number, if the predications stop downwards as well as upwards.

Having proposed to prove that the predications cannot proceed to infinity and that there are immediate premises (which we require for the sciences) he divided predications into those that are natural and those that are unnatural, so that he might begin by specifying that our present account concerns natural predications, not unnatural ones. And he posited for us three hypotheses concerning them: for either, when the last subject is determinate, the predications go infinitely upwards,[41] or, when the last predicate is determinate, the advance downwards is infinite,[42] or when both of the extremes (both the subject and the predicate) are determinate, the intermediates are infinite,[43] and it is precisely this that he now proposes to refute first.
For the fourth hypothesis, I mean, that which posits an advance to 225,1
infinity both downwards and upwards, is refuted along with the first two, since if neither is true, by itself, clearly they will not [be true] together. He proves that, when the extremes are determinate, the intermediates cannot be infinite, in the following manner. He takes A to be the predicate, F to be the subject, and B to be the items between them. So if when A and F are determinate, the intermediates, for instance, B, are infinite, one who starts from A, can never reach F; nor can one [starting] from F [reach] A. For the infinite cannot be traversed. But if that were the case, the extremes would not have been limited. For the advance of things starts from the most universal things and proceeds to those that are most particular, but by nature our knowledge [goes] in the opposite direction, starting from the most particular things and proceeding in this manner through the middles to those that are most universal.[44] So it is clear that it is entirely necessary that the intermediates are finite in number. For nature would never reach what is most particular, nor

would our knowledge [reach] the most universal, if there were an infinite number [of items] between them. For if, whenever we proceeded from more particular things to ones that are more general and encompassing, we knew that A is the most universal, it would clearly follow that we would never know that A is the most universal, since we would not have travelled through all of the middles and would not have found that[45] which encompasses them all. For we do not know whether [an item] through which we did not travel is more universal than A nor whether it is encompassed by it, and similarly [we do not know] whether it is more universal or particular than F. So when the extremes are determinate it is entirely necessary that the intermediates are not infinite in number. So saying this[46] is just like saying that while the number [of numbers] from one to ten is limited at the extremes, one and ten, [the items] between them are infinite. So just as this is not possible (for the number will be infinite in actuality, not by virtue of being [infinitely] generated), likewise, the intermediates cannot be infinite when the extreme terms are determinate. But if someone were to say 'Why can't it be the case that, just as the extremes of a continuum are determined, (even though, since every continuum is indeed infinitely divisible, the items between them are infinite), so in the case of predications, the extremes are determined but the items between them are infinite?'[47] we will respond to this that in the case of continua the infinitely many intermediates are not already subsisting in actuality, but are potentially infinite by virtue of infinite division, but in the case of the predications it is entirely necessary that the middles already
226,1 subsist in actuality. Even if they are not taken, they must nevertheless exist, if they are classified up to some most generic [term] A.[48] Well then, if the extremes are determinate, and the intermediates are infinite in number, just as I said, nothing follows but that, while there is a finite number in the direction of the smallest determinate number and in the direction of the greatest [determinate number], the numbers intermediate between the greatest and the smallest are infinite in number,[49] which is impossible, since, for the total number [of terms], when the extremes are determinate, the intermediate numbers too are surely finite. Therefore when the extremes are determinate the intermediate terms cannot be infinite in number.

> **82a30** For even if someone were to say that of ABC, some are next to each other with the result that there are no intermediates, while the others cannot be taken, that makes no difference, ...

[He says this] since it has been proposed to prove that demonstrations do not proceed to infinity but come to immediate premises,

which do not require demonstration. And it is not possible to prove them through demonstration; rather, they are self-evident.[50] And for this reason he proposed to prove that the predications do not proceed to infinity. Now a hypothesis is put forward, which says that the extremes are determinate but that the intermediates are infinite in number. Now, since he said that [on that hypothesis] we would never go from one extreme to the other (for we must always travel through [predications] that are mediated), in order that one might never say that it has been incorrectly hypothesised that the intermediate predications are mediated (for some are immediate, as well) – for this reason he says that, if not all of them are mediated but some are immediate and some are mediated, the same thing will nonetheless result. For if we travel from [the point] at which there are mediated premises towards the other extreme, we will never reach it. For example, if between F and A there were BDCE, and we were to start from A and come upon immediates up until C, and then from C on, the remaining premises were mediated, it is clear that we would never reach E from C. Nor, for the same reasons, will we ever reach A from E. Therefore [on this hypothesis] the intermediates must be infinite in number. For even when the advance is through immediate premises, those who start from one or the other of the extremes might never reach the rest. For if it should happen that G is predicated adjacently of F, and B has a position adjacent to A, and so on, we will never arrive at the highest items, in a situation in which that which is under A is predicated ad- 227,1
jacently of something in the direction of F, because we hypothesised that the middle terms are infinite in number, and the infinite cannot be traversed.

> **82a32** ... for whichever of the Bs I should take, the intermediates in the direction of A or F will be infinite.[51] [For it makes no difference which is the first from which they are infinite in number, whether or not this happens at once. For those after them are infinite in number.]

He says that when the middles are infinite, whichever of them is taken, the items that are intermediate between it and the one or another extreme, either A or F, must be infinite in number. So again, in the case of those that are infinite[52] <...> we will never be able to reach [it]. For if neither those in the direction of F nor those in the direction of A are infinite in number, it is entirely necessary that the whole be finite, but the middles are supposed to be infinite in number. So it makes no difference whether or not the advance through immediates happens at once.

Chapter 21

> **82a36** It is clear also that it will stop in the case of the privative demonstration, if in the case of the affirmative one [it comes to a stop in both directions].

In the text before us he proves that, since affirmative demonstrations do not go to infinity but stop, it is entirely necessary that negative
15 demonstrations too stop and that they not go forward to infinity. This can be made clear by a single argument. For if every negative thesis is proven when an affirmative premise is taken, and there can be no syllogism without an affirmative premise, and affirmative premises come to a stop, it is, I suppose, entirely necessary that negative premises too come to a stop and that there are some immediate ones.
20 For if they did not stop, but there is always another item to which, primarily, it does not belong, that one will surely be affirmatively joined to that before it, so, if the negations proceed to infinity, the affirmations will proceed [to infinity] along with them. But this is impossible; therefore the negations too will stop. He methodically
25 proves this for the three figures. For since the negative thesis is proven through the three figures, he goes through each of the figures and proves that in none of the figures can negative demonstrations proceed to infinity.

> **82a38** For let it be possible neither to go upwards to infinity from what is last [(I mean by 'last' that which itself belongs to nothing else, but to which something else, for example F, [belongs]) nor [to go] from what is first to the last (I mean by 'first' that which is itself [said] of something else, but of which nothing else [is said]). If these things are so, they will stop in the case of negation, too.]

30 He says, let it be assumed by us by hypothesis – in order that you might take what has been proven as being hypothetical[53] – that it is impossible to go affirmatively to infinity from the last item to the first through the intermediates and [it is impossible to do this] from the first to the last. For he does not say 'let this have been proven', but
228,1 let what has not yet proven be assumed by us as proven.[54] I mean [by 'what is not yet proven'] that what starts from below does not go to infinity <but> reaches some last item which is only a predicate, and similarly that which starts from above does not go to infinity but comes to some last item which is only a subject.[55] There are still two hypotheses
5 that entail infinitudes; these have not yet been refuted. He now considers it to have been granted to him that this [infinitude] is impossible, in order to on that basis prove that negations cannot go to infinity, for it is after this that he takes up the argument concerning affirmations and

proves through several [arguments] that it is impossible for affirmations to go to infinity in any way whatsoever.

> **82b4** There are three ways to prove that something does not belong.

That is, it is proven in [all] three figures that something can be denied of something, but in every instance one must arrive at immediate negations. First, he taught which are immediate negations.

> **82b5** For either B belongs to everything to which C belongs, and A belongs to nothing to which B [belongs] …

First, he proves that it is impossible for the negations to go to infinity in the first figure. For let A belong to no B, and let B belong to every C. It is clear that A will belong to no C.[56] It is also clear that, if A is denied of B not immediately but through some other [term], for example, D, D must belong to every B, for in this way A will be proven to not belong to B through the middle term D. For A belongs to no D, D belongs to every B, and A to none of the Bs. Again, if A is denied of D not immediately but it is first [denied] of something else, for example, of E, E, again, will necessarily belong to every D, and so forth. So if it is impossible for A to be denied immediately but whenever anything has been taken there is always something prior to which it does not belong, then, if the negations proceed to infinity, it is entirely necessary that the affirmations too proceed to infinity. But it has been assumed that that is impossible. So denials too cannot proceed to infinity. For not always will there be something else of which the major [term] is first denied, should that [middle term] be indeed predicated affirmatively of the minor [term], seeing that the major [term] is denied of the minor through the middle.

> **82b6** So in the case of BC and every instance of the second interval, one must make one's way to immediates. [For this interval is affirmative. Clearly if the other one does not belong to something else that is prior, for example, to D, this will necessarily belong to all B. And again, if it does not belong to something else prior to D, this [other one] will necessarily belong to every D.]

That is, the minor premise, which is affirmative, must not proceed to infinity on account of how it has been assumed by us that the affirmations do not proceed to infinity. 'Make one's way to immediates': that is to say that at some point as they go forward the affirmations reach an immediate premise, with the result that nothing else more universal than those terms can be found.

> **82b11** So since the path downwards stops, the path upwards, too, will stop,[57] and there will be something primary to which nothing else will belong.

He calls 'the path downward' the path to the minor term, that is, the [sequence of] affirmative premises.[58] For since, he says, when they are predicated of the minor [term], it is not possible to take such middle terms to infinity, but they stop when the most universal [term], of which it is not possible to affirm anything else, has been taken, clearly the path upwards too, that is, the denials, will stop. For it is an immediate denial, when the major term is denied of the most universal genus. But there must be such a middle term. It follows that if the affirmation stops the denial too will stop.

> **82b13** Again, if B belongs to all A and to no C, A belongs to no C. [Again, if one must prove this, clearly it will be proven either in the above manner or in this [manner] or in the third [manner].]

Having proven for the first figure that it is not possible for negations to proceed to infinity, he now proves [this] in the second [figure] as well. So again there is a middle term B, denied of the minor [term] C, and affirmed of the major [term] A, so that the major premise is
230,1 affirmative. So he concludes that A [belongs] to none of the Cs.[59] So if B is denied of C not immediately but through another [term], for example D, it is, he says, possible to deny B of C through the middle D in the first, second, and third figures. We already showed how [this is so] in the first [figure]. But now we should say [how this is so] in the second [figure]. For if D belongs to every B and to no C, one will conclude that B belongs to none of the Cs. And again if D is denied of C not immediately but through the middle E, again, when a major premise ED is generated, as well as the minor, negative [premise] EC, one will conclude that D belongs to none of the Cs. And in the second figure one must always make the major [premise] affirmative, and the minor, negative. For the sake of clarity let us work through the argument with terms. To begin, let horse be instead of the major term A, and instead of the minor term C, stone, and instead of the middle B, neighing. So neighing belongs to every horse, but to no stone. And it follows that horse will belong to no stone. If neighing is denied of stone not immediately but through another middle [term], for example, perceptive, perceptive belongs to every neighing thing but to no stone, and thus neighing will belong to no stone. Again if perceptive is denied of stone not immediately but through the middle term animal, animal will belong to every perceptive thing but to no stone, and accordingly perceptive [will belong] to no stone. Again if animal is not denied immediately of stone, etc. So if we proceed in this way and the affirmations necessarily stop, it is entirely neces-

sary that the denials too stop. Now it is also possible to prove the same thing if the minor premise is taken to be affirmative,[60] sometimes [increasing] the minor and sometimes the major,[61] but if we need to continuously increase the affirmations, then whichever affirmative premise we initially took, we must also subsequently take what follows.[62] We will take things in this way if we always take middle terms that are more universal than the initial middle term.[63] For when we sometimes take them in this way, but sometimes take ones that are more universal than what is denied,[64] we produce a syllogism in this manner, but do not continuously increase the affirmations. For example neighing [belongs] to every horse and to no stone, and horse belongs to no stone. So in proving that neighing belongs to no stone I am able to have a proof, if we take as middle term that which is more universal than neighing, for example, perceptive, in the very manner that we produced [the syllogism]. But I 231,1
am also able to take what is more universal than stone, for example, imperceptive (for this [belongs] to every stone but to no neighing thing) and we can conclude that neighing belongs to no stone. But formerly I continuously increased the affirmation; for neighing belongs to every horse, and perceptive to neighing and animal to this. However in the latter case, this is not how [it is], but sometimes the affirmation is at the major premise and sometimes it is at the minor premise.

> **82b17** One might in this way prove, for example, that D belongs to every B [but to no C, if something must belong to B].

That is, it is possible to prove the negative premise through the middle figure, if the middle term is situated in such a way that something else is predicated of it, that is, if it is not the most universal, but D is predicated of it.[65] For it is clear that it must be denied of C and likewise B will be denied of C through a middle. The order of the argument is reversed. What follows [proceeds] in the same way. One might use <such> a proof, if something must belong to B; an example is the case in which D belongs to every B but to no C.

> **82b19** And again, if this will not belong to C, [something else belongs to D, which does not belong to C.]

He says that if 'this', D, is not denied of C immediately, there must be something else, for example E, that belongs to every D but to no C. For in this way D will belong to no C as well.

> **82b20** Therefore since belonging to what is higher always stops, [not belonging too will come to a stop].

He already said what upwards and downwards and first and last mean for him, that he calls 'upwards' and 'first' what is itself predicated of another, but of which nothing else [is predicated], and similarly [he calls] downwards and last what is itself predicated of nothing, but of which something else [is predicated].[66] So he says that, since affirmations stop upwards, that is, [since] they stop once they have reached what is most universal, the negations too must stop, if it is impossible for a mediated negation to be proven without an affirmation.

232,1 **82b21** The third way was: if A belongs to every B, and C does not belong, C does not belong to every A. [This, again, will be proven through the ways discussed above, or in a similar manner. Now [if it is proven] in the one way they will come to a stop, but if in the other way, again it will be assumed that B belongs to E, to every one of which C does not belong. And this again, likewise, and since it has been assumed that it comes to a stop downwards, too, clearly C's not belonging will also stop.]

He moved on to the third figure. Nothing universal can be proven in this [figure]; rather, all [conclusions] are particular. Yet he admits that in this figure too the universal negation [is proven] for the sake of completion and accordingly he proves that it is not possible for negations to go to infinity in this figure either.[67] He takes the minor [premise] AB to be affirmative (for in every instance of the third [figure] the minor [premise] is affirmative), and the major [premise], CB, to be a denial. So if C [belongs] to none of the Bs, and A to every B, C does not belong to some A.[68] Now if one needed to prove through another middle term that C belongs to no B, again, in the third figure one must always make the major [premise] a denial and one must always take a middle term that is more particular than the minor, so that the minor can be predicated of it. For example, let the major [term] be animal, the minor [term] imperceptive, and the middle term lifeless. Now imperceptive [belongs] to every lifeless thing, and animal to no lifeless thing, and animal does not belong to any imperceptive thing (according to the premises). So if one needed to prove the denial, I mean that animal [belongs] to no lifeless thing, he must take a middle term that is more particular than the minor [term], I mean, than lifeless thing, for example, stone: lifeless [belongs] to every stone, animal to no stone, and animal to no lifeless thing.[69] Again, if one needed to prove the denial, he must take a middle term that is more particular than stone, for example, say, magnetic stone, and so forth. Now since the affirmations come to a stop downwards, it is clear that the denials too will come to a stop. And it is clear that in the first and second figures the path is to that which is more universal because the middle term is of wider extent than the minor, but here in the third [figure] it proceeds downwards, since the middle term is more particular than the minor [term].

> **82b29** Clearly, even if it is proven not in one but in all ways, sometimes by the first figure, sometimes by the second or third, it will in this way too come to a stop …

Since he made the demonstrations in each of the figures separately, he says that even if the syllogisms proceed not in a single figure but, say, the negative premise is proven in the first figure, and the negative premise that is used in its proof is not also proven in the first 233,1
figure but in the second, and likewise again the negative [premise] that is used in its proof [is proven] in the third, and so forth, and in this manner the proofs of the negative premises are varied in their figures, even so they must reach immediates. For if all of the figures are three in number, and in each [of them] both negative and affirmative premises have been proven to be finite in number, clearly what is composed of all [of them] will be finite in number.

> **82b32** [… for the paths are finite in number, and all finite [numbers] multiplied by a finite [number][70] must be finite. [So it is clear that they come to a stand in the case of privation, if they do so in the case of belonging too.]

He says 'finite [numbers] multiplied by a finite [number]' because, as I said, both the [number of] figures and the [number of] premises in each of them are finite. Some of the manuscripts have 'a finite number of times' instead of 'multiplied by a finite number'.[71] He adds this, because, even if there were a finite [number] of premises in each [syllogism], there could still be an infinite [number] of them all, if the number of figures[72] were infinite. But [as things are] now, all are three in number.

Chapter 22[73]

> **82b35** That [this is true] in these cases, too, is apparent for those who consider it formally, as follows. Now this is clear in respect to the predicates in the what it is. For if the essence can be defined or is knowable, and the infinite cannot be traversed, the predicates in the what it is must be infinite.

Having proven that when the extremes are finite in number the number of middle [terms], too, must be finite, and that when the universal affirmations are finite in number, the negations too must be finite, he moves on to the remaining hypotheses, I mean that the advance neither upwards nor downwards is infinite. He pays special attention to proving that one cannot go upwards to infinity. For this has special need for being proven. For it is fairly obvious to everyone that predications do not go downwards to infinity. For who does not know that among predications, the indivisibles are last, and that these are predicated of nothing?[74] But whether or not it is possible to ascend upwards to infinity, and up to what point, is not easy to see. He proves this through several arguments, first, in a way that is more formal, and then also in a way that is more concrete. Alexander[75] says that the arguments on the basis of [the nature of] definitions, [the arguments that] he first employed,[76] are more formal.[77] For he assumes that there are definitions and that it is possible to define things, but does not prove it. He assumes that there are definitions, as something agreed upon, just as in the second of the formal arguments he assumes that there is [such a thing as] demonstration. However the Philosopher[78] says that it is not for this reason that [Aristotle] calls the arguments formal. For unless we introduce the view that there is no 'apprehension',[79] it is evident that there are definitions.[80] But as things are, we know in what respect a horse differs from human being and in what respect it has anything in common, and likewise in regard to the rest. This comes from knowing the common essential [features] of each as well as the properties that belong to it, which are the basis for definitions.[81] But again, he did not say that Aristotle called 'formal', [arguments] that are convincing but not true[82] but those that are from true premises and that are [themselves] true, yet are not demonstrative but are more common and can apply to a number of things, by means of which he proves not only that per se predicates do not proceed to infinity, but proves that no predicates at all do.[83] The proof from definitions, which he first employs,[84] is of this sort. For, he says, it made it clear that those predicates of a thing that are in its what it is cannot be predicated to infinity. For example, rational animal is predicated in the what it is of human being. But again, substance, living, perceptive are predicated in the what it is of animal. That it is not possible for things to proceed in this way to infinity, he proves as follows. If we do know things, and each is known through its own definition, but definitions come from the thing's genera and its own differentiae, then surely, if we know the definitions, we must know each of the things that are included in the definition, again including both the genus and the differentia of each thing by means of each thing's own definition. If then the predications ascend to infinity and everything that has been taken has something else more universal and generic than itself, but no one can know anything unless he knew both what is its genus and what are its differentiae – and it is impossible to traverse an infinitude – it follows that it is impossible to define or know anything in any way at all. For in order to know human being, one must grasp its definition, that is, its genus and differentiae, and in order to know each of these, one must do the same for them too, and so on to infinity. If then it is impossible to traverse an infinitude, it is therefore impossible to know or define human being. So if this is false and we do define things, and we define [them] truly, it is therefore impossible for the predications to proceed to infinity. Now he called this argument 'formal', first, because the taking of definitions belongs to a formal method,[85] and second, because by the same [argument] we prove that no other [kind of] predication proceeds to infinity. For we will say the same things about quality and quantity. For the knowledge of these too [comes about] through definitions. For this reason the essential predications of accidents do not go to infinity. For if each of the categories[86] is finite in number, it is clear that what is composed of all of them, too, will be finite.[87]

> **83a1** We argue universally like this: for it is possible to truly say that the pale walks and that that large is a stick [and again that the stick is large and that the human being walks.]

Since he wants to prove that predications are not infinitely generated, a division is made among predicates, in regard to in how many ways they are predicated. He already did precisely this.[88] But now he fully works through the division in a more complete manner. For since among beings there are these two, substances and accidents,[89] either a substance is predicated of a substance or an accident of a substance or a substance of an accident or an accident of an accident. An example of a substance [predicated] of a substance is when we say that a human being is an animal. An example of accident of a substance, is when we say that a human being is pale. And he calls these predications in the strict sense. For a more general substance is naturally predicated of a more particular substance, and further, so is an accident of substance. For an accident subsists in substance as a subject. But when substance is predicated of an accident, as when we say 'that white is a stick' or an accident of an accident, as when we say 'that bald is pale', such things, he says, either should not be called predications at all or this whole thing is an accidental predication and an unnatural predication. For an accident does not naturally serve as subject for an accident, nor is an accident [naturally a subject] for a substance.[90] He divides these things too, and calls the predication that predicates an accident of an accident, an accidental predication in the more proper sense (for it happens[91] to the bald that it is also pale), but that which predicates a substance of an accident he calls unnatural, for that which predicates that which is naturally a subject and posits it as a predicate is unnatural in the strict sense [of the term]. It is possible to naturally predicate an accident of an accident, as when we predicate what is more universal of what is more particular, for example, 'white is a colour'.[92] Further, perhaps in these cases you should not even say that an accident is predicated of an accident. For being a colour is not an accident of white, but white is a colour, since the more universal item is predicated of the more particular. [This is] why one might not say that an accident is predicated of an accident.[93] Therefore we should make a twofold division of this part and say that when an accident is predicated of an accident, either the predication is unnatural, as when an accident from one category is predicated of [one from] another, or it is natural, as when both are from the same [category].[94] So since predications are said in this many ways, we now investigate whether the natural predications, which are also predications in the strict sense, go on to infinity or not. And it will have been likewise proven as an additional result that it is not possible for the unnatural predications to go on to infinity either.

'We argue universally like this, for it is possible to truly say that the pale walks.' 'Universally' is instead of 'in general': we should say in how many ways every predication is said. To say that the pale walks is to predicate an accident of an accident. He did not mention this before. [To say] that that big is a stick is [to predicate] a substance of an accident. For 'big' belongs to quantity, and stick belongs to substance. Again 'the stick is big' is [to predicate] an accident of a substance. He now passed over [the case of] predicating a substance of a substance, for example, 'human being is animal' since he already discussed it in the lines in which he just said 'Now this is clear with respect to the predicates in the what it is.' [Clearly] he passed it over on account of its being obvious.

83a4 Speaking in this way or that way are different.

'This way', as when we say that the stick is big; 'that way' as when we say that that big is a stick, etc. The one is natural, the other, unnatural.

237,1 **83a4** For when I say that the white is a stick, I then say that that for which being white is an accident is a stick [but not that the white is the subject for the stick].

That is, I do not say that that white itself is the subject for the stick (for that is impossible) but that this, the substance of which the white is an accident, is a stick.

83a7 For it is not the case that, insofar as it is white, or insofar as it is precisely a particular white, something became a stick, [so that it is not [white] except accidentally.]

The subjects for natural predicates are either precisely[95] what each predicate is or precisely what [each] particular predicate [is].[96] (For example, human being is said to be an animal, and a human being is either precisely animal or precisely a particular animal. For when the subject and the predicate are coextensive, then what is precisely the predicate is also the subject, for example, a human being is receptive of intellect and science. Human being[97] is precisely what is receptive of intellect and science.[98] The same thing [holds] in the case of individuals, too. For example, the approaching is said to be a human being. And it is either precisely a human being or is precisely a particular human being. And similarly this is white, and either is precisely white or is precisely a particular white). So for this reason he says that, when I say that this particular white is a stick, I am not saying that what is precisely a white or what is precisely a particular white[99] is a subject for a stick (for the quality does not by nature serve as subject for the substance) but that that to which being white is an accident is a stick. If this is so, we accordingly say that the white is accidentally a stick, and [the predication] is unnatural in a rather strict sense, because the natural order is reversed.

> **83a9** But when I say that the stick is white, [I am] not [saying] that something else is white, and that being a stick is an accident of that, [as when I say that the cultured is pale (for then I am saying that the human being, of whom being cultured is an accident, is pale) but the stick is the subject, and it is precisely that which became [white],[100] not being something different from what is precisely a stick or a particular stick.]

He says that [things are] not [as follows]: as when we say that the cultured is pale, we say that something else serves as subject for the pale, which is precisely what happens to be cultured,[101] and for this reason we also say that the cultured is pale accidentally – likewise when we say that the stick is white, we say that something else is a subject for the white, to which being a stick is an accident. Rather, we say that precisely what is stick, or precisely what is a particular 238,1
stick, is white. For the very substance of the stick serves as subject for the white.

> **83a14** Now if we must lay down a rule, let speaking in this way be predicating [and let speaking in the other way be either not predicating at all, or predicating not in the strict sense, but predicating accidentally. But the predicate is like white, that of which it is predicated is like stick.]

The ancients did not think about positing names for the distinction among these sorts of predicates, so for this reason he assigns names, the very thing that he did in the *Categories* in the discussion concerning relations, when he named something 'with a rudder' and 'winged' and 'headed' and the like.[102]

> **83a18** So let it be posited that a predicate is always predicated in the strict sense, not accidentally, of that of which it is predicated. For this is how demonstrations demonstrate.

Having distinguished in how many ways predicates are predicated, and having said 'Now if we must lay down a rule', let this be predicating in the strict sense, and that, 'either not predicating at all or doing so accidentally', he fully works through the present [material] in more detail. He takes up the argument and fully works through these things one point at a time. So he says that it should be posited that all of those items that are not predicated accidentally are predicates in the strict sense. But he added 'always', not because it is necessary that a predicate be always predicated, but because [it is] always [the case that] all of those [predicates] that are not predicated accidentally are predicates in the strict sense.[103] He has already said which are predicated in the what it is of substances or are accidents of substances. That these are predicates in the strict sense, he proves in the [lines] in which he says 'for this is how demonstrations demonstrate'.

> **83a21** So [whenever one thing is predicated of one thing] it is either in the what it is or it is [predicated] that it is a quality or a quantity etc. [or a relation or action or passion or place or time.]

If demonstrations demonstrate the items that belong to things, that is, either the what it is of things, I mean, the definitions, or the quality of substances, or the quantity or one of the other categories (for example, when we investigate concerning heaven, say, whether it is composed of the four elements or from some other substance,[104] we investigate the what it is of it; and when [we investigate] whether it has a spherical shape or what other sort of shape it has, the quality; and [when we investigate] whether it is infinite or finite,[105] the quantity, and when [we investigate] whether it encompasses all beings within it or not,[106] the relation, and when [we investigate] whether or not it is active on the things here, and if it is active, whether it is also acted on in turn or not, action and passion, and whether or not it is eternal,[107] the time, and likewise in regard to the other [categories]); so if all demonstration is about these things and they are predicated of substances, I mean the definitions and their parts, and the other categories, then it seems that these are the ones that are predicated in the strict sense. That there are only so many categories that are predicated of substances, he grasps from induction. For there is no doubt that, of everything that is, whatever you might mention is classified under one of these. Well then, doesn't the geometer demonstrate accidents of shapes, for example, of triangle, that the three angles are equal to two right [angles], and of circle, that [the lines] from the centre are equal, and doesn't the arithmetician [demonstrate] accidents of number? How then is it that they are not demonstrating accidents of accidents?[108] We say that these things, even if they belong accidentally to shapes, nonetheless form an essential part of their substance and are as it were differentiae constitutive of species, by which they are separated from the others.[109] So those who are considering these things are not considering the things that belong accidentally to shapes but their very subsistence, how it is and from which things they are composed. For just as receptive of intellect and scientific understanding or mortal or any of the things in its definition do not belong to human being as one thing does to another – rather, they form essential parts of it – so too the circle is studied on the basis of all things that are seen within it, and likewise too the triangle, since that which does not have the three angles equal to two right angles or the two sides greater than the remaining ones would not be a triangle, but if one of these were to be removed, [its] being a triangle would be destroyed at once. And this is how it is in all cases. Except that this is obvious,[110] that no science demonstrates either accidents of accidents, for example, say, paleness of three cubits tall, or a substance of an accident, as stick of white. So it is clear that such predications are unnatural and accidental. Someone might be puzzled concerning how it can be that he talks about demonstrations that prove accidents of substances, for example, that the earth has a spherical shape or that it has these qualities, cold areas, say, or dry areas.[111] And we say that the natural scientist proves that the earth is spherical in the following manner, not as though such a shape belonged per se and primitively to the essence of the earth, but as proving that sphericity belongs strictly speaking to the shape of the earth.[112] Even if there were another earth and the same things were accidents of this as well as that, it doesn't matter; he nonetheless proved that sphericity belongs to that too, for sphericity belongs per se to the shape of the earth.[113] So it is either for this reason that he says that [instances of] scientific understanding concerned with the accidents of substances are demonstrations,[114] or, what is more likely, the term 'demonstration' seems to be used here in a more general sense and not according to its [senses] previously distinguished by him, so that we should understand 'demonstration' in the general sense here as every true proof, that is, which is from true premises, just as we say that the public speakers and grammarians demonstrate in the general sense, should they demonstrate something truly.

> **83a24** Further, [let it be posited that] those [terms] that signify substance are precisely that or signify precisely that particular of which they are predicated.

One must understand 'let it be posited' as common [to both clauses]. So <let it be posited> by us, he says, and let it also be agreed that, when something is essentially predicated of something, that of which it is predicated is either precisely the predicate or is precisely that particular [predicate].[115] We already discussed this. For if, when I aim at a definitional account, I say that Socrates is a human being or an animal, that which is precisely the predicate is also the subject;[116] however, if I should consider the predicate as a genus or species, Socrates is precisely a particular animal or precisely a particular human being.[117] The text is somewhat badly put. He says 'Further, [let it be posited that] those [terms] that signify substance', that is, all of the predicates that signify substance, 'are precisely that or signify precisely that particular of which it is predicated'. He seems to say that the predicate is precisely the subject; he means to say the opposite, that the subject is precisely the predicate.[118] Now this is how we should construct the text: further, the items that signify the substance of that of which they are predicated[119] signify either precisely that item, that is, the predicate, or signify precisely a particular [instance of] that.[120]

> **83a25** All of those items that do not signify a substance but are said of some other subject, which is neither precisely that item nor precisely that particular item, are accidents [for example, the pale of the human being. For a human being is neither precisely pale nor precisely a particular pale thing ...]

He says, let this too be posited by us, that all of those things that are said of some other subject are called accidents, and their subject is neither precisely the predicate nor precisely the particular predicate.[121] For when I say that the human being is pale, the human being is neither precisely pale nor precisely a particular pale.[122] This is why the pale is an accident of the human being, as it does not complete it in respect to its being.

> **83a29** But surely [he *is* an] animal, for a human being is precisely an animal.

That is, a human being is precisely an animal. He adds 'surely', not because [the matter is] in dispute, but since he is not (now) concerned with mentioning what belongs essentially to human being.

83a30 All of the items that do not signify substance must be predicated of some subject, and there cannot be something white which is white while not being something different.

And let this be posited, he says, that all of those items that do not signify substance have their being in a subject: for as is his wont he here employs 'of a subject' instead of 'in a subject'.[123] For none of the accidents subsists itself per se,[124] but there must first be something else, such as human being or stick or some such thing, and then there is white or black or three cubits or some such thing. Then, since the argument concerning the Ideas clashes with this position[125] – (For they say that there are Ideas for all things, which transcend all matter and subsist by themselves, which very things they refer to as 'the thing itself', for example 'animal itself' and 'human being itself' and 'equal itself' and 'beautiful itself', and 'precisely animal' and 242,1
'precisely human being', 'precisely equal' and 'precisely beautiful', and so forth. The reason why they refer to them like that, as being purely what they are in fact called, is this: the beautiful and equal *here* are not precisely the beautiful and precisely the equal, but are mixed with the ugly and the unequal, I mean, [they are mixed] with matter. For it [i.e. matter] is ugly insofar as it is without form, which is why there is no pure form here, since it is mixed with the formless. And the equals [here] are not equal in the strict sense. For they might sometimes come to be unequal too. But the *logos*[126] of equality will never accept the opposite, in addition.) – now, since these [points] are opposed to the present position which is to the effect that there is no essential form separate from matter, for this reason, he speaks in a way that attacks the doctrine of the Ideas.

83a32 So goodbye to the Forms. For they are but noodling and even if there were such things, they would be irrelevant in regard to the argument. For demonstrations are concerned with these sorts of things.

The inarticulate preliminary fingerings[127] of the kithara players that occur for the sake of testing the sound of the strings are called noodlings, so [saying] this is as if he said 'the arguments concerning the Ideas consist in only empty words,[128] devoid of sense'. For how can there subsist a whiteness or humanity or equality itself by itself? For since all of these forms and those like them are material,[129] they cannot subsist otherwise than in some subject, I mean in a body or in absolute matter.[130] How is it that there subsist in themselves the things that by nature do not exist per se, when, he says, even if they exist, they will in no way stand in opposition to our present argument? For demonstrations are concerned with forms of a certain kind, those which cannot subsist apart from matter, for geometry and

arithmetic and all the rest concern forms of that sort, those that are in matter. So we will not go wrong if we say that these things do not admit of subsisting in themselves but entirely have their being in a subject. And when they speak in support of them, they say that
243,1 Aristotle himself explicitly proclaims on every occasion that the demiurgic accounts of things are Ideas.[131] For it is he who in the *Metaphysics* says that just as the order[ly layout] in a camp does not arise spontaneously, but from the order within the commander, so the order in the cosmos does not [arise] spontaneously but from that in the Demiurge.[132] And <...> that in the doctor who is ill.[133] And it is he himself who says that the demiurgic intellect sees all things when it sees itself,[134] and that intellect insofar as it is a plenitude of forms, is also a form. Further, he says in *On the Soul* 'those who say that the soul is the place of forms speak rightly'[135] but, they say, it is in regard to those who misunderstand the doctrines concerning the Ideas and think that whiteness subsists by itself and not in the demiurgic *logos*, or [think this of] a bodiless humanity, as if it had nose and feet and hands and such things, that he was always wont to attack the argument about such Ideas.[136] But I find thoroughly unconvincing the defence to the effect that even though it was Plato who posited the Forms as demiurgic *logoi* existing within the Demiurge, Aristotle always himself says the same things and never objects to this, and that, even if he meant the same thing [as Plato], while Plato says that Ideas are such a thing,[137] others misunderstood him.[138] But this is not what Aristotle is up to, here. Rather, it is obvious that he is always doing battle against the doctrine [of Forms], not those who conceive of it incorrectly. For in the *Metaphysics*[139] he draws out many long refutations of the doctrine. It is related that even while Plato was alive Aristotle directed against him very powerful refutations of this doctrine. And it is obvious that in reality Plato did not say that Forms are simply *logoi* within the Demiurge, but he granted them subsistence in themselves and [said] that the equal itself and the animal itself and such things were something, and also said that it is in regard to these, as paradigms of images, that the Demiurge fabricated the things here.[140]

244,1 **83a36** Another point is that if this is neither a quality of that nor that of this, and there is no quality of a quality, they cannot be mutually predicated in this way …

Since he said that some predications are natural and some are unnatural, he now wants to specifically prove that such predications are in reality unnatural. At the same time he proves that unnatural predications cannot proceed to infinity either. He proves this through a so-called objection and counter-objection.[141] First he made use of the counter-objection, and then the objection. For having first conceded that there are unnatural predications, he proves that such [predications] cannot proceed to infinity;[142] then [he proves] that unnatural predications cannot be predications in the strict sense.[143] So he first works through his argument as applied to one category, quality, and then also as applied to all unnatural predications in general.[144] These are all of those that either predicate a substance of an accident or an accident of an accident, for example, pale of three cubits tall, or cultured of snub, except for the cases in which the predicate belongs in the what it is of the subject. Cases like this are those in which [the predicates] are genera of the subjects, for example, colour of white, or their species, (for example, white in general, of this white, I mean, the individual one) or their differentiae (for example, dispersive, of white,[145] and rational or mortal of human being). And even if we concede that unnatural predications are predications, it is clear from the following that not even predications of this kind can go to infinity. For let it be assumed that stick is predicated of white. Stick is either predicated of white in the what it is or as an accident of it. But if it is predicated in the what it is, the predications cannot proceed to infinity. For we proved this (that the things predicated in the what it is are finite) on the basis of the [nature of] definitions.[146] If they are predicated accidentally, again, they too are similarly finite, for the predications that proceed from accidents go forward to the items that are predicated in their what it is.[147] For example, if pale is predicated of Socrates, as it proceeds the predication will predicate colour of paleness and quality of this. These are predicated in the what it is (as are, for example, quality of colour, and this of pale). So, again, since [the items] predicated in the what it is are finite, such predications, too, must be finite. But he shows as follows that these are not predications in the strict sense. First, as I said, he develops the argument in respect to one of the categories; his example is quality. He begins by assuming a hypothesis, that the same thing – a quality – cannot be of itself;[148] rather, there is no quality of a quality in the strict sense. For example, if whiteness is a quality, it will not involve something else which very thing would be as it were a quality of the pale. You will say the same thing in respect to every [category], that it is not possible for the same thing to be a genus of the same thing,[149] or for the same thing to be a quality of the same thing.[150] For example, if white is a quality of a stick, stick cannot be a quality of a white. And if animal is a genus of a human being, human being cannot also be the genus of an animal. This is known from its being evident. So if this assumption is made in the beginning, let one of the categories, quality, be taken by us, and let species that are not subordinate, such as snub and pale, be predicated of one another. So if snub is not predicated in the what it is of pale (for they are not subordinates) clearly, snub is a quality of pale. But pale too is a quality. Therefore there will be a quality of a quality, which is impossible. And since not only is snubness predicated of pale but pale [is predicated] of snubness (for we say that this snub is pale) pale will therefore be a quality of snubness. Therefore each will be a quality of the other, which is impossible. For this same reason there will not be a quantity of a quantity nor will any of the other [categories] be [predicated] of [something of] that same [category]. For the same reason, a substance cannot be predicated of one of the accidents in the what it is [of the accident], a point on which we have already agreed.[151] For if stick is predicated of white in the what it is, as, for example, a differentia or a genus, and again white of stick, clearly stick belongs essentially in white. But since white is also predicated of stick (we grasp this because it is evident) it must also be the case that white is predicated essentially of stick. For if two things are mutually predicated, the way in which one is predicated of another, whether essentially or accidentally, is the same way in which the other will have been predicated of the first. For example, if philosophical were predicated of Socrates, clearly philosophical would be an accident of Socrates. So even if someone were to say that this philosophical [one] is Socrates, clearly philosophical is an accident of Socrates. Similarly, if sweet is an accident of this wine, it is also the case that being wine would be an accident of this sweet [stuff]. And if the philosophical one is accidentally bald, it is also the case that the bald is accidentally philosophical. This is how things are for essential predications too. For example, if the animal is essentially living, the living will essentially be an animal. And the same holds for all [cases]. Therefore if someone were to say that stick essentially belongs to white, since it is also the case that white is predicated of stick, it is also the case that white will have been essentially predicated of stick. But if both are mutually essentially predicated, he says, something will itself be precisely what it is. For he said that the subject is precisely the predicate or is precisely the particular [predicate] itself, when the predicate is essential. If then the stick is precisely white and the white is precisely a stick, it is also the case that the stick is precisely a stick and also that the white is precisely white, and the same thing will have been predicated of itself. But this is ridiculous. For in this way the same thing will be a species or differentia of itself. For essential predicates are either genera or species or differentiae of the subjects. If then stick is the genus of white and again the white is a genus or differentia of stick, you will infer that stick is the genus or differentia of stick, therefore it is precisely itself. So if these [results] are impossible, it follows that it is impossible for either substance to be predicated of accidents or qualities of different kinds to be predicated of each other.

Perhaps one might raise a puzzle that by this argument it is not possible for certain items to be mutually predicated; yet is true to say both that animal is essentially living and that what is living is

[essentially] an animal. For if the genus of animal is living, and a certain living thing is an animal, it follows that something itself is precisely itself and is its own genus. So perhaps one must say in such cases too,[152] that while it is true to say that a certain living thing is an animal, still this sort of thing is not a predication in the same way. For the predicate must be one thing and the subject another. But the particular living thing is the same as animal. Therefore one thing is not predicated of another. So this sort of thing, 'the sword is a long knife' or 'the human being is mortal', is not a predication in the strict sense; rather, while it is possible to speak truly in this manner, it is no longer a predication.[153] Likewise in the case of properties, should we say that a human being is capable of laughter, it is a predication, for we say that the accident belongs to the substance, but if we say that the one capable of laughter is a human being, we say what is true, but such is not a predication, for it is not possible for capable of laughter to be a subject for human being. It is clear from this that in general it is not possible for a predication to be unnatural, for none of the accidents can subsist in themselves but all of them have their being in something else; it is therefore impossible to make the accident a subject and to predicate something else of it, unless it is predicated in its what it is, as colour is of white. For colour is the genus of white, and colour is not predicated of white as being in a subject. So if I say that the pale is snub, clearly I am saying that the 247,1
snub is in the pale as being in a subject. But this is impossible. Therefore unnatural predication cannot be a predication.

> **83a38** But while it can be truly said, it is not possible for them to be truly mutually predicated.

While it is true to say that this white thing is a stick and that this snub [one] is philosophical, it is not true to call such a predication, for he already said that the things predicated in a natural manner are predicated in the strict sense, but the things [predicated] in an unnatural manner are either not predicated at all or are predicated accidentally. So it is not the same to say that something is [the case] and to predicate.[154]

> **83a39** For either it will have been predicated as a substance [for example, as being either a genus or a differentia of the predicate. But it has been proven that these will not be infinite, either downwards or upwards …]

One must add 'or as an accident'.[155] This is what permits us to understand it [in this way]: he added [this point] in the development of the argument.[156] For after he showed how it is not possible to predicate [something] as a substance he went on to say 'Nor will any [instance] of quality or any of the others, if it is not predicated accidentally.'[157] Neither the essential nor the accidental predicates are infinite either upwards or downwards.

> **83b4** For example, a human being is two footed, and this is an animal, and this something else.

Because the predicates surely come to a stop upwards when they reach the most generic genera.

> **83b4** And there is no [infinitude if one predicates] animal of human being, and this of Callias [and this of something else in the what it is].

Because they also come to a stop downwards. For individuals are predicated of nothing else.

248,1

> **83b5** For every such substance can be defined, but it is not possible to pass through an infinitude [of things] by conceiving [of them]. [So they are infinite neither upwards nor downwards, for it is not possible for that substance of which infinite items are predicated to be defined.]

He calls 'such' not the most generic substance but every one after that, for example all of the subordinate genera and species up to the individuals. For it is not at all possible for the substance of the most general items to be defined, but it is known from description.[158] Therefore if every such substance can be defined, the predications cannot proceed to infinity. For the one who is defining must go through all of them, in order to know each of the things that are included in the definition. But it is impossible to pass through an infinitude [of things]. So if we are truly defining, the predications are not infinite in number. He added 'by conceiving' instead of 'by knowing each of the things taken in the definition'.[159] For it cannot be defined in any other way. One who renders a definition 'conceives of' each [part] of each thing. So if the predications [proceed] to infinity, those who define a substance must go through its predicates to infinity. But this is impossible.

> **83b9** So genera will not be mutually predicated. For then something would itself be precisely the particular that it itself is …

After first having conceded that [there is] an unnatural predicate that is predicated as a genus or differentia and having proven that the predications do not [proceed] to infinity, he now makes an objection against this very point,[160] to the effect that things cannot be mutually predicated as genera. We have [already] said enough about this.

83b10 ... nor will any [instance] of quality or any of the others, if it is not predicated accidentally. [For all of these are accidents and are predicated of substances.]

Having said 'for it either will be predicated as a substance' and having through omission left it to us to understand 'accidentally' and then having argued against the [case in which it is predicated] as a substance, he now argues against the case in which the unnatural predicate is predicated as an accident. He first developed his argument in regard to one category, quality, and then said 'another point is that if this is not a quality of that'. But now he proves by a general argument that none of the unnatural predications can predicate something accidentally of that of which [something] is predicated. For neither is substance [predicated] of accidents; nor are the others[161] predicated of one another. The demonstration is that all of the categories are accidents of substances, and none are accidents of the others, because nothing can subsist in other things, unless it is in substance.

83b12 But [we must prove] that they will not be infinite in number upwards ...

He hypothesised three hypotheses according to which it is possible for predications to go to infinity. He refuted one, that is, the one that posits that when there are determinate extremes the intermediates between them are infinite,[162] and then having proven that in general when there is a finite number of affirmations it is entirely necessary that the denials be finite in number and not go to infinity,[163] he moved on to the remaining hypotheses and also proved that [predications] cannot proceed to infinity upwards. First he proved that the predicates in the what it is are not infinite.[164] The genera and differentiae of species are predicated in the what it is not only in the case of substance but also in the case of the other categories. He proved that the predicates in the what it is are not infinite through a reduction to absurdity, for, he says, if they were infinite, things could be neither defined nor known. For should we want to define human being and if we were to say it is a mortal rational animal, in order for us to truly know what the definition means, we would have to define each of the things included in the definition, and, again, [define] each [of the things included in the definition] of *them*. So if the predications do not come to a stop, it would be impossible for anything to be defined, for the infinite cannot be traversed, and it is impossible to know something when its principles, that is, [the principles] of the genus, cannot be determinately known. For example it is not possible to know human being if animal is not known, and again, we do not know what 'animal' means if its genus is not known, and this does not [go on] to infinity; likewise in regard to quality and quantity and the others. Now having said these things and wanting the argument to extend not only to substantial predications but also to the accidents of substances (this is because these too are predicated of substances, for we say that a human being is pale or three cubits) he took from division the number of ways in which predicates are predicated. And he showed that some predicates, those that are naturally predicated, are said to be predicated in the strict sense (and all of the items that are predicated in the what it is are naturally predicated, such as animal of human being or colour of white), as well as all of those items that are in that subject of which they are predicated, as accidents of substances. Unnaturally predicated are both a substance of accidents and accidents of accidents (all of the items that are not predicated in the what it is). Then he showed that unnatural predications do not go on to infinity either.[165] Having proven these things, he turns back to the matter before him, which was to prove that accidents are not predicated of substances infinitely upwards either. He again shows this by means of examples. This is why we said that such a proof is called formal:[166] because it applies not only to items that belong essentially and per se but also to all of those that are naturally predicated in any way whatsoever. Still, one cannot use those arguments in regard to all predicates whatsoever, which is why he then proves demonstratively and not formally that predications do not go to infinity. He thus calls 'formal' the proofs that are more general and apply to a number of things. In this passage he proves in the way I explained that the predication of accidents of substances cannot go to infinity. He proves [it] on the basis of [matters] that have been proven. For if it has been proven that all of the predicates in the what it is are finite in number, and that the kinds of predications are finite as well – for we grasp this because it is evident, because everything that you might say is classified under one of the ten categories – then if the kinds of predications are finite in number, and the genera or species in each are finite in number (for this has been proven), clearly there is also a finite number of essential predications of accidents, for that which consists of finite [constituents] is itself finite, too. Perhaps it would not be a bad idea to work through the argument with an example. For example, if pale were predicated of human being, and we wanted to know what in the world this is that has been predicated of human being, this must surely be defined as a colour dispersive of sight. So we will predicate colour of pale, and again, if you please, quality of this. It has been proven that all of the predicates in the what it is are finite in number, and likewise for quantity and the rest. It has therefore been proven that the essential predications of accidents upwards are finite in number. That the predications downwards, that is, to the subjects, are finite as well, he proves in a summary fashion. For, he says, all accidents are predicated of substance. For all of the items in the essence (upwards, the predicates in its what it is, and downwards, the individuals) are finite in number. For it is obvious from its being evident, that individuals are predicated of nothing else. So if essence is finite going downwards, and both all items in the what it is and all accidents of substance are predicated of substance, it is clear that all of the predicates are finite in number, both upwards (for this has been shown from definitions) and downwards. For of all of the predicates the last is the individual substance. For the very last [predicate] is a substance. At any rate, every accident is predicated of a substance, and the predications of accidents leave off at this last item, I mean the substance. Of substance, the individual substances are the last, for these, as I said, cannot be predicated of other things. But it is obvious that the predications of substance too leave off at the individuals. Therefore it has been proven that predications are finite in number downwards, too, for down from every predication there are individual substances.

'But it is also [clear] that they will not be infinite in number upwards.' It is clear that predicates will not be infinite in number upwards.

> **83b13** For there is predicated of each of these what it means or what sort of thing it is or how much it is or one of such things[167] [or the items in the substance. These are finite in number, and the kinds of predicates are finite, for they are quality or quantity or relation or action or passion or where or when.]

Then he enumerates what kinds of predicates there are, [noting] that they are either the categories of accident or the items in the substance. He says that the items in the substance are those predicated in the what it is of substances. These are the genera and species and differentiae. All of these, he says, have been proven to be finite in number upwards. Then, since, even if each category encompasses a finite [number], there can nonetheless be an infinite number of predications, if the kinds of predications are not only ten but are infinite in number (for if there are an infinite number of kinds, and all are predicated of substance, even if each encompasses a finite [number], it will again happen that the items predicated of substance are infinite in number), and this is why he also makes this additional point, that the kinds of predications are finite in number. He accordingly enumerates them.

83b17 It is assumed[168] that one thing is predicated of one thing …

That is to say, in demonstrations. For he also said this previously, namely that demonstrations prove some one of these, quantity or quality or one of the others: 'whenever one thing is predicated of one thing'. It is possible for compound predications to be made out of multiple ones, as when I say 'Socrates is a philosophical human being' but the predicated term is not a unity. For in syllogisms there must be simple terms. He made this additional point, not because, if one thing is not predicated of one thing, [the thesis] that predications are not infinite is somehow voided, but because it is predications of this sort that need to be used for syllogisms.

252,1 **83b18** And these things, which are not what it is, are not predicated of themselves.

That is, this too is assumed by us, that no category is itself predicated of itself, if it is not predicated in the what it is, for example, pale is not predicated of bald, [even] if that is how things turn out. However colour is predicated of pale, for colour is the what it is of pale.[169] But one accident is not predicated of another accident. He takes up these matters now, so that he might distinguish unnatural from the natural predications.

83b19 For all are accidents, though some are per se and some otherwise [but we say that all are predicated of some subject, but an accident is not a kind of subject …]

For colour belongs per se to pale, but to Socrates otherwise. For it is an accident of him and does not belong to him per se.[170] The conjunction 'for' [indicates] the giving of the explanation for what has come before, I mean, of [the fact] that accidents are not predicated of each other, if only one thing can be the subject for accidents.

83b22 For we posit nothing of this kind to be that which it is said to be while it is not something different from what it is said to be …]

He proves through these [remarks] that none of the accidents can be a subject for something else. For each of them is said to be precisely what it is said to be, while being something else that is prior.[171] For the white is either a stick or stone or some such thing, and is then called white. The case is similar with three cubits tall, or right-side, or any of the others.

83b23 But it is itself [said] of another thing, and some other things are predicated of [yet] something else.[172]

He says that each of the accidents does not itself serve as subject for anything. But it is said of something else, and another thing of another thing, for example white, if you will, of snow or white lead, and black of pitch and ebony or some such thing. Some of the manuscripts have 'it is itself [said] of another thing and this, of another'. [On this reading] he would be saying, for example, that if pale is predicated of human being, and this of Socrates, pale is also [predicated] of Socrates.

83b24 Therefore neither upwards [nor downwards will one thing [be said to belong] to one thing.] 253,1

*** or is predicated of several items,[173] through these [arguments] he proved that it is not possible for one thing to be predicated of one thing to infinity. In the words that follow,[174] he will show that [it is also] not [possible] for one thing to be predicated of several things [to infinity]. For never are many things taken to be predicated of one thing within a syllogism.[175] [The conclusion] also [follows] in a different way. If one thing is not predicated of one thing to infinity,[176] it is clear that several things are not predicated of one, either. For within several things there are many units.[177]

83b26 The things of which accidents are said are all of those that are in the essence of each; but these are not infinite.

Having said that there is not an infinitude either upwards or downwards, he proves each of these [points]. First [he explains] how [there is not an infinitude] downwards. For he says, the things of which accidents are predicated, that is, the subjects of accidents, are substances. Substances are genera, species and individuals. For we say that the animal is pale, and that the human being is pale, and that Socrates is pale. But substances are finite in number. For we said that the individuals leave off at the last item. So if substances underlie every predication, it follows that these predications are finite downwards.

83b27 Upwards there are both these and the accidents; neither of these is infinite.

[He says that] they are not infinite upwards. For of predications, he says, those that are upwards are those that are in the what it is. For we say 'a human being is an animal, an animal is living, what is living is a body, a body is a substance'. And all of the accidents are finite in number, for it is these that are predicated of substances. As indicated above,[178] all of these are finite in number, and at that time we proved that all of the predicates in the what it is are finite in number. For it was then proven by appeal to definitions that each of the predications is finite in number. So both upwards and downwards the predications are finite.

254,1 **83b28** Therefore there must be something of which something is first predicated [and something else of this, and this must come to a stop and there must be something which is no longer predicated of anything prior and that has nothing prior predicated of it. So this is said to be one manner of demonstration.]

With these [remarks] he concludes the three hypotheses.[179] There must, he says, be some last item of which something is primarily predicated, and this is the individual, of which the most specific species is primarily predicated. And when the predications proceed like this continuously they come to a stop at one last item of which nothing else is predicated. So the extremes and the middle terms are finite in number.

83b32 But there is yet another [manner], if there will be[180] a demonstration of those things of which certain prior things are demonstrated, etc., [and for those things of which there is demonstration, it is not possible to be better disposed towards these things than by knowing them, nor is it possible to know [them] without demonstration.]

He calls this manner [of demonstration] formal, not for the same reason [as before][181] (for the things that he says here do not apply to more than one [kind of matter] but only to those that are demonstrable) but because he assumes that there is [such a thing as] demonstration. Here there might well apply the interpretation that Alexander gave in regard to the previous argument, when he said that [the argument] is formal for this reason – that it assumed, as a matter of agreement, that there are definitions.[182] Yet he proved at the start[183] that there is [such a thing as] demonstration, for which reason, this does not seem to have been assumed without demonstration. So one must not interpret 'formally' as applying to both [lines of argument][184] because he does not work through the argument that he produces in what follows, in respect to per se predications themselves, but as regarding certain other [matters], for example, the assumptions that there are definitions and that there are demonstrations. But in the present argument he proves that when the extremes are determinate the intermediate predications cannot be infinite. Now after he here assumes as a matter of agreement that there is [such a thing as] demonstration, he again argues by a reduction to absurdity that if predications go to infinity, demonstration is eliminated. He assumed as a matter of agreement that there is [such a thing as] demonstration because this was proven at the start in the [passages] in which he raised objections against those who eliminated demonstration and against those who said that all [objects of scientific understanding] are demonstrable.[185] The proof goes like this. He says that there is demonstration of those things for which there are certain prior things. The things of which there is demonstration cannot be known except by demonstration, for demonstration derives its conviction in secondary things from prior things. It is impossible to know the things of which there are prior things, if those prior things are not known in advance. For there are two ways of knowing, one from demonstration, which derives its conviction in secondary things from prior things, and that which is superior[186] to that which is by demonstration, I mean, [the way] of [knowing] common notions,[187] by which we know immediate premises and common notions, not knowing them on the basis of prior items (for there are none prior to them) but by apprehending them immediately, as sensation [apprehends] sensibles[188] (while it is impossible to know demonstrable items in this way). So if the progress of predications towards the middle term is to infinity, an immediate premise cannot be found. But if there is no immediate premise – rather, for every one that is taken there can be another intermediate term – demonstration is impossible. If demonstration derives its conviction of secondary items from certain primary items,[189] the secondary items cannot be known if the primary items are not known. But we know the primary items either, again, by demonstration, should there be other items prior to them, or in a manner superior to demonstration, should they be common notions. But if what is taken has something preceding it, and so on to infinity, then, since it is not possible to traverse an infinitude and since there is no [form of knowing] superior to demonstrative [knowledge], it would necessarily follow that one does not know it at all. So the ones who say that predications proceed to infinity eliminate demonstration and knowledge in general as well as apprehension.[190] Since this is absurd and there is demonstration, predications do not proceed to infinity. One can prove not only that the intermediates are not infinite but that the items upwards are not either.[191] For if these are infinite, demonstration is eliminated as well, since demonstration does indeed [occur] through definitions, and, as we earlier proved, it is not possible to render definitions if the predications upwards are infinite.

83b35 If this one is known through those, and we neither know them nor have a better grasp of them than by knowing, we will not have scientific understanding of that which is knowable through these.

The secondary things, of which there are certain items that are prior and have the character of being a principle to a higher degree, cannot be known if the primary things are not known. For he says that the primary things are either themselves known through demonstration, if they have something prior to them, or [are known] in a manner superior to demonstration, if there is nothing preceding them. But, he says, if the primary things are not known, on account of the infinite advance of predications, the things with which one is acquainted on the basis of the primary items cannot be known. [The words] 'we neither know them' [are used] instead of 'we neither know them by demonstration'. He contrasted 'nor have a better grasp of these than by demonstration' to this.[192]

83b38 So if there is such a thing as knowing by demonstration, in the strict sense and not on the basis of certain items or by hypothesis, [the intermediate predications must come to a stop. For if they do not come to a stop, but there is always something above what has been taken, there will be demonstration of all things.]

I said that the reason that it has been posited that there is demon-
256,1 stration is that he demonstrated it at the start. 'Not on the basis of certain items or by hypothesis' is [to say] 'If not every demonstration proceeds by hypothesis but there is such a thing as demonstration in the strict sense that does not employ any hypothesis.'[193] A demonstration that assumes something demonstrable and by that means proves something else is called a demonstration by hypothesis,[194] for example, if someone were to say 'let what I am saying be granted: that there is providence, and I prove that the soul is immortal'[195] or 'let it be granted what I am saying, that the soul is immortal, and I prove that it is separate from bodies' for such demonstrations are by hypothesis and are not [demonstrations] in the strict sense.

84a2 So if it is not possible to traverse[196] an infinitude, we will not know by demonstration the things of which there is demonstration. [So if we are not better disposed towards these things than by knowing [them], there will not be scientific understanding through demonstration in the strict sense, but by hypothesis. One might in a formal manner derive conviction concerning what has been asserted from these [considerations] ...]

If, he says, there is something above anything that has been taken, all things must be demonstrable, for demonstrations are on the basis of prior items. But since an infinitude cannot be traversed, again it will not be possible to demonstrate anything, since it is impossible to demonstrate when no principle is known. So the argument comes around to a contradiction, that all things are demonstrable and that nothing can be demonstrated.

> **84a8** It is clear analytically [in a more concise fashion] from these [considerations][: that in the demonstrative sciences, about which the investigation is concerned, there cannot be an infinite number of predicates either upwards or downwards.]

Having proven in general that no predication proceeds to infinity, he now intends to prove the same thing specifically in respect to demonstrables. There are two ways in which demonstrables are per se. For they are either those items that are included in the definition of the subjects, as predicates in the what it is, such as animal of human being,[197] or they are things that include their subjects in their definition, as odd and snubness and the like.[198] Now he proves that none of these proceed to infinity. For example animal is predicated per se of human being, and living of this, etc., [and he proves] that these cannot go to infinity; similarly odd belongs per se to number, and prime to this and something else, say, to this, [and he proves] that these cannot proceed to infinity either. He calls such proofs 'analytic' instead of 'demonstrative' because the aim of the whole analytic enterprise is demonstration. This is why he wrote at the start of the *Analytics* 'First one must say about what and of what there is the inquiry, that it concerns demonstration and is of demonstrative science.'[199] Since he accordingly called the whole enterprise concerning demonstration analytic, naming the whole on the basis of the part, for this reason he called all demonstrative proofs analytic. For throughout the whole method our aim is to investigate the ancients' syllogisms, [to see] whether they are necessary and demonstrative, and to both skilfully analyse them and reduce them to the figures, and to investigate whether their matter is demonstrative and necessary. This is why the whole enterprise was named from its most useful part. He proves that the predicates in demonstrations are not infinite either, in the following manner. For items in demonstration are predicated per se. As we saw, there are two ways in which these [are per se]. First, there are those that are included in the definition of the subject, for animal and rational belong per se to human being, [and] similarly lifeless to stone, and likewise in other cases. All of those things that include the subject in their definition are per se in the second manner, as odd and even [take] number, and straight and curved [take] line, and snub [takes] nose. And he first proves that the things that are per se in the second manner do not proceed to infinity. For example, if odd belongs to number, it is surely necessary that number belong in the definition of odd. Again, if <prime belongs to> odd, both odd and number will belong <in the> definition[200] of prime, and similarly if something else will belong per se to prime, all of the lower items, prime and odd and number, will be included in the definition of that. For since the present argument concerns per se items, and of those, [per se items] of the second kind, and the hypothesis has these proceeding to infinity, clearly the subjects will always be included in their definitions, not only the proximate ones but also those that are farther. For if prime belongs to odd, surely it is necessary that not only odd but also number be included in the definition of prime.[201] So if the per se items are infinite, and these belong in actuality, it is clear that the things included in their definitions will be infinite.

But it seems that Aristotle has not reduced the argument to this [absurdity], for somebody might say against it that it will not be possible to take the last of the things that belong per se, if they are infinite, and if this is so, what is included will surely be from the intermediates, and the items included in its definition will no longer be infinite. For the subject, such as number, is defined. So whichever of the things that belong per se I take, surely the intermediates between it and number, which [intermediates] must be included in its definition, will be finite in number.[202] The reason why this reduction to absurdity does not seem to be necessary, is that the predicates are not included in the definition of the subject, but the other way around. Now in truth the absurd [conclusion] is appropriately applied to the second way of being per se, because the predicates are included in the definition of the subject,[203] and it was assumed that these are infinite. So it was to be expected that Aristotle does not reduce the argument to this absurdity,[204] but [says] that if there were an infinite [number] of things that belong per se, it would happen that an infinitude will belong to one thing, in this example, to number. For prime, which belongs to odd, will clearly also belong to number. And should something else belong to prime, this same thing will belong to both odd and number. So it will happen that an infinite [number] will belong to one thing, in this example, to number. But if this is impossible, he says, it is clear that they will not be infinite upwards, which was assumed. He calls 'upwards' from number to odd and prime etc. Therefore the things that belong per se in the second way cannot proceed to infinity.

So Aristotle reduced the argument to this [absurdity] when he said that an infinite [number of things] of that kind cannot belong to one thing. It is clear that [things of] that kind of thing are the things that belong per se. He made this additional point, not because it is possible for an infinite [number] of other items that do not belong per se to be in one thing but either because the present argument concerns things that belong per se, and he said what happens to the [things that are] per se, or [he said this] in order to make a contrast to the parts of a continuum, because they are infinite in potentiality. He said simply this much, that an infinite [number of things] cannot belong to one thing, but he did not add [an explanation of] why it is not possible. But Alexander says that it is because if this were to be the case, there results an infinitude in actuality. But it does not seem appropriate to introduce this into [the text] before us, for he supposed [that] this very thing [was the case], that 'there results an actual infinity'. So this will seem to be nothing absurd. Now the Philosopher[205] said that the argument bears on definitions, because it will result [from the assumptions] that an infinite [number of things] will be included in definitions. But this also does not seem to me to be true; I mean, Aristotle's argument is not this reduction to absurdity.[206] For he says that an infinite [number] belongs to one thing, the subject, as we will show when we go through the text. The things that belong to it per se are included in the definition of number. So his argument bears on definitions not because it is not possible for them to be composed of an infinite [number of things], but because if the things that belong were infinite in number, and they belong per se to it, it is impossible to know <it> if one does not know all of the things that belong per se to it (for we could not know the nature of triangle if we did not know what are the things that belong per se to it); so if there is an infinite number of items that belong per se to things, and it is impossible to know and traverse an infinite [number], it is therefore impossible to have knowledge of things.[207] So if this is absurd, an infinite [number] of per se items cannot belong to one thing. In addition to these points, Alexander makes another one, when he proves that there cannot[208] be an infinite [number of things] that belong per se in the second manner. For these, he says, always proceed to the lesser.[209] For just as odd is lesser than number and prime is lesser than odd, likewise, if something else belonged to prime, that would be lesser than prime, and those that next proceed in like manner to the lesser would [eventually] terminate in individuals. Therefore such things cannot proceed to infinity. For if there were an advance to what is greater, the following might make sense, that it could proceed to infinity, as does the progress of number to what is greater. But if there is an advance to the lesser, it must surely reach individuals, as the progression of number towards the lesser reaches the monad. Therefore it is not possible for the items that belong per se to things in the second way to proceed to infinity.

This is how [he argues] concerning the second way in which [things are predicated] per se. He easily proves from definitions that things that are per se in the first way, too, are not infinite in number. For if these are included in the what it is of their subjects, as animal [is taken] in the definition of human being and living in the definition of animal, and this proceeds to infinity, it is not possible to define. So if this is false and there is knowledge through definitions, the items predicated of things in the what it is are finite in number.

84a11 For demonstration is of all of the things that belong per se to things. [There are two ways in which things are per se.]

The article *tôn* modifies 'all of the things that belong per se to things'; there is demonstration of them.

84a13 Both all of those items that inhere in[210] them in the what it is …

This is the first way. He says that all of those items are said to be per
260,1 se, which, in those items <to which> they are said to belong per se, belong to them in the what it is, for example animal is in the what it is of human being.

84a13 … and those things for which the things to which [they belong] belong in the what it is for them.

The second way. And those items, he says, are also said to be per se, to which the very items to which they are said to belong per se, that is, their subjects, belong to them in the what it is. For example <…> nose belongs per se to snub, it belongs in the what it is of snub itself.

84a14 For example odd to number [for odd belongs to number, and number itself belongs in its definition, and again multitude or divisible belong in the definition of number.]

In this [passage] he gives examples of the two ways: odd, of the second, (for this belongs per se to number, and number is included in the definition of odd) and number and divisible, of the first, for these things, which belong to number, are included in its definition, for whenever we define number we at one point included multitude (for we say that a number is a multitude collected out of monads) and at another point include divisible, saying that number is a divided quantity.

84a17 Neither of these can be infinite, nor [can those that are predicated] as odd of number …

If we understand 'neither' [to refer] to both multitude and divisible, both of which are examples of the first way, it was to be expected that he add 'nor as odd of number', so that he might indicate the second way, as well. But if 'neither of the two ways' is meant, 'nor as odd of number' is said by way of repetition, since he proposed to refute this first.

84a19 For, again, there would be something else in odd,[211] to which item, which belongs, it inheres ...

He says that if something else will belong per se to odd, to which odd will necessarily belong in its account, while it belongs per se to odd, number too must be included in its account. Now the Philosopher,[212] <who> inserted a comma after 'but if this is' so that it means 'if this 261,1 is how things are and something else belongs to odd' read the rest continuously from 'first'. If this is how things are, he says, and something else also belongs to odd, number or odd will be the first to inhere in this too, and it is the first for one who starts from above, as if we were to say that substance or animal is the first to belong to human being. But it does not seem to me that the comma is rightly placed [by the Philosopher]. For Aristotle everywhere[213] calls first the [items that are] adjacent, not those that are farthest. So the comma must be placed after 'prime'.[214]

84a20 But if this is prime,[215] number will inhere in what belongs to it.

Having said 'there would be something else in odd' he tells us by example what sort of other thing [there might be]. For, he says, if this should be the case, let us posit that prime, for example, belongs to odd. He calls prime a number that is measured by the unit alone as a common measure. This belongs to no even numbers outside of two alone. For it alone among even numbers is measured only by the monad. However it belongs to many odds,[216] 3, 5, 7, and countless others. This is why one might say that prime belongs to odd per se. So, he says, if prime will belong per se to odd, and number, like odd, will be included in the definition of prime,[217] [prime] too [will be] among the items that themselves belong to number.

84a21 So if there cannot be an infinite [number] of such things belonging to[218] a single thing ...

He says 'such things'; that is, the per se predicates.

84a22 ... they will not be infinite in number upwards either. [But all must belong to the first, for example, to number, and number to these ...]

When he says 'upwards' he means from number, both odd and prime and so forth. If things were infinite like that, clearly an infinite [number of items] belongs per se to some one thing, number. So if this is impossible (for in this way we would eliminate scientific understanding), the advance from number through an infinitude of things that belong per se to it is impossible.

262,1 **84a24** … so they will be [terms] that convert and do not extend further …

After he said that, if an infinite [number of items] cannot belong (on account of how an infinite [number of items] cannot belong to a single thing) they will not be infinite upwards, either, he thereupon developed this point. For if all of them necessarily belonged to number and if number [necessarily belonged] to all of them, clearly both would convert and number would not extend farther and be of wider extent than all of the things that belong per se; nor would the things that belong per se extend farther than number.[219] But if this (for an infinite [number of items] to belong to number) is not possible, neither is it possible for number to belong to an infinite [number of things].

84a25 Nor are all of the items that belong in the what it is infinite in number. [For then it would not be possible to define. So if all of the predicates are said per se, and these are not infinite, they come to a stop upwards, as they also [come to a stop] downwards.]

It follows that for [predicates] that are per se in the previous way, too, the predications will not progress to infinity.

84a30 But if this is so, clearly there must be principles of demonstrations and there is not demonstration of all things [which as we said at the start, is what some people say. For if there are principles, not everything is demonstrable and it is not possible to proceed to infinity.]

Having proven that it is in no way possible for predications to progress to infinity, he still has to draw his conclusion, for the sake of which he initiated all of these [remarks], I mean, that immediate premises are principles of demonstration, and that demonstrations cannot proceed to infinity. For if premises were not immediate and for this reason demonstrations did not come to a stop, but for every premise that is taken it is possible to insert a middle term to make a syllogism, it would follow that when there are two determinate terms, there could be an infinite [number] of intermediate ones, which has been proven to be impossible.

84a33 This is because for either of these [options] to be the case is nothing other than there to be no immediate and indivisible

> interval [but for all things to be divisible. For it is by internally inserting a term, as opposed to taking one in addition, that what is demonstrated is demonstrated, so if this can go to infinity, there could be an infinite number of middle terms between two terms. But this is impossible if predications come to a stop both upwards and downwards. Earlier it has been proven formally that they come to a stop, and now [it has been proven] analytically.]

To say that demonstrations make their way to infinity and that there are no principles of demonstration or that all [matters] are demonstrable is, he says, to say nothing other than to say that there is no immediate premise. For demonstrations proceed not by taking an additional term from outside,[220] but [by inserting them] in the middle, for example if A belongs to B and B to C, A will belong to C. So if we want to prove the source of the fact that A belongs to B, we will insert a term D intermediate between A and B. Similarly, if we want to prove that A belongs to D, we again will insert an intermediate term E, and so forth. So between two intermediate terms there will be an infinite number of ones placed within them, which is impossible. For it has been proven that the predications that begin from above come to a stop downwards, and [those that begin] from below [come to a stop] upwards, so if, when there are two terms, it is impossible for there to be an infinite number of intermediates, neither are all things demonstrable (for immediate premises are indemonstrable) nor does demonstration proceed to infinity, but immediate premises are principles of demonstration.

Chapter 23

> **84b3** Now that these things have been proven, it is clear that if the same thing belongs to two things, for example, if A [belongs] to both C and D …

Having proven that immediate premises are principles of demonstration, he infers from these [considerations] another theorem, that when some one item is predicated of several items, and none of the others is in every case predicated of the other it is not always the case that the predicate is predicated of them by virtue of some common [item], but there are times when it is predicated immediately. What is an example of what I am talking about? Substance is predicated of Socrates and Alcibiades by virtue of some common item, living, say, for, since both Socrates and Alcibiades are living, and what is living is a substance, substance is predicated of Socrates and Alcibiades because of this common item, living.[221] Similarly, living is predicated by virtue of some common item, animal, and similarly animal is predicated of them by virtue of some common [item], rational. So then is this how all [such cases] are predicated? But if all of them [are predicated] in this way, it will not be possible to take an immediate premise, but, for every premise that is taken, one could add an intermediate term. So there will be an infinite [number of terms] between any two intermediate terms, which is impossible. For human being is predicated of them, and not by virtue of a common thing; rather [it is predicated] of both of them immediately. He added this theorem since prior to this he hypothesised that one thing is predicated of one thing and in this manner he proved that it is not in this way possible to proceed to infinity. He added these [remarks] in order to show that it is not possible to proceed to infinity not only when one thing is predicated of one thing but also when one [is predicated] of several.

264,1 **84b4** … and if one of them is not predicated of the other, either not at all, or not in every case, they will not always belong in virtue of something common.

He needed to add this: that the subjects must be mutually predicated either not at all – in which case they are either individuals or are distinguished from one another, as substance <might be predicated of body>[222] through the middle term living, and might be predicated of animal through this middle term, and might be predicated of rational and irrational through this middle, animal (for neither of these is predicated of the other since they are distinguished from one another) – or if one is indeed predicated of the other, it is not predicated in every case. For example, suppose that the rational were also the mortal. For mortal is not predicated of everything rational; nor is rational [predicated] of everything mortal. As I said, it was necessary for him to add this. For if one thing is predicated of every instance of another, for example, if the rational were also a human being (for rational [is predicated of] every human being), surely the predicate is not still predicated of both by virtue of something common, as for example, say, substance [is predicated] because of animal, but it is predicated of the one because of the other, for it[223] will be predicated of human being because of the middle term rational. It is not predicated of both through yet some other thing. So he was right to add 'and one of them is not predicated of the other, either not at all, or not in every case'. Clearly, this same [point] will apply even if those of which it is predicated are not two but more [than two].

84b6 For example, [this is so] if having [angles] equal to two right angles belongs to scalene and isosceles by virtue of something common, for it belongs insofar as it is a certain shape, and not insofar as it is something else …

Some of the manuscripts have 'to two right angles', and some 'to four'. So if it is 'to two right angles', and as some manuscripts have it, it is conjoined with 'for it belongs insofar as it is a triangle',[224] the meaning is clear, for having two right angles belongs to isosceles and scalene not insofar as it is isosceles (for this is the manner by which it has equal [angles] at the base, but not being equal to two right angles), and not insofar as it is scalene, but insofar as each is a triangle. So having [angles] equal to two right angles belongs to them by virtue of something common, triangle. But if [the reading] should be 'to 265,1
shape' one must put the circumflex on the penultimate of '*schêma*' so that it is in the nominative case and the '*ti*' must be taken as an enclitic so that it is an indefinite pronoun, for one might say that because [they are] a certain shape, that is, a triangle, the [attribute] that is mentioned [i.e. two right angles] belongs to them.[225] But if it should be 'to four right [angles]', the '*schêmati*' must be taken to be a proparoxytone since it is dative, and the account concerns the exterior angles.[226] For it is proven that the exterior angles of every shape, when its sides are extended, equal four right [angles]. And the proof of this I have expounded in *Miscellaneous Theorems*.[227]

> **84b9** But this is not always so. [For let B be that in virtue of which A belongs to C and D. Clearly B also belongs to C and D by virtue of something else that is common, and that by virtue of something else, so that between two terms there would fall an infinite [number] of terms. But that is impossible. So the same thing does not necessarily belong to many by virtue of something common, if there will indeed be immediate intervals.]

That is, the predicate will not always be [predicated] on account of something common to them, but for those that proceed there will be that which will have been predicated immediately of them.

> **84b14** However the terms must be in the same genus and derived from the same individuals, if the common thing is to be among those things that belong per se. [For as we saw, the [terms] that are proven [to belong] cannot cross from one genus into another.]

This refers to the demonstrative terms, since dialectical [terms], at any rate need not belong to the same genus; this is because dialectical proofs are also on the basis of reputable [premises] and accidental [predications]. However, demonstrative premises must be of the same genus. He says 'of the same genus' instead of 'of the same subject'[228] in order that the predicate be classified as within a genus, since it is not possible to prove by crossing from one genus into

another. For it was said above that it is not possible to demonstrate medical [matters] geometrically and grammatical matters rhetorically. For this reason it seems to me that by 'of the same genus' he does not simply mean the items under the same category. For more than one science deals with the same category. Rather, by 'genus' he means that which belongs to a science. For example [he says] that the predicate and subject terms in geometry must fall under the classes [found] in geometry, for example, if you will, under lines or shapes or some such thing, and likewise in the case of the other sciences. 'Derived from the same individuals' means 'predicated of the same individuals' so that the individuals too, of which the terms
266,1 are predicated,[229] are classified under the same genera and the terms in the syllogism, both the predicate and the subject, are also predicated of the same individuals, either affirmatively or negatively.

> **84b19** It is clear that, when A belongs to B, if there is a middle [term], it is possible to prove that A belongs to B. [Also, the elements of this are the same as and as many as the middle [terms]. For the immediate premises – either all of them or the universal ones – are elements. But if there is not, there is no longer a demonstration; rather, this is the way to the principles.]

Since it has been proven that it is possible to immediately predicate one thing of another, he says that it is clear that when something is predicated of another in this way, there is no demonstration of it, if every demonstration comes about through the insertion of a middle term. So it is not possible to demonstrate the sorts of [conclusions] for which it is not possible to posit a middle term, for immediate premises are not demonstrable – on the contrary, immediate premises are principles and elements of demonstration. Definitions are primary principles, but among premises, unmediated premises are principles and elements.[230] Now, he says, all immediate premises, or the universal ones, are principles and elements of syllogisms. What sort of thing do I mean? In order to demonstrate that a human being is living, I employ this sort of syllogism: a human being is rational, what is rational is living, therefore a human being is living. Such premises are not immediate. For the major [premise] is demonstrated[231] by having animal taken as a middle term. For what is rational is an animal, what is animal is living; therefore what is rational is living. And 'an animal is living' is an immediate premise.[232] This is how things are in the case of the minor [premise], too, when animal and rational are placed as middle terms. Now, he says, all immediate premises, or the universal ones, are principles and elements of syllogisms. By 'universals' he means either the common notions[233] or those that are the most generic, for example, say, body is a substance, or some such thing.

> **84b24** Likewise, if A does not belong to B, too, [if there is a middle term or a prior [term] to which it does not belong, there is a demonstration, but if not, there is not.]

[He says] that, just as in the case of affirmations those that are mediated are demonstrable, while the immediates are not demonstrable but are principles and elements of demonstration, this is how things are in the case of negations too. We grasp as clearly demonstrated that not everything is demonstrable, if demonstrations are on the basis of principles, and immediate premises are [such] principles, and there are no principles of immediates. For it has been proven that there are immediate premises.

> **84b26** But there are as many principles and elements as there 267,1
> are terms.[234] For the premises that belong to them are principles of demonstration.

For all of the terms that are immediately predicated of each other, too, are principles, but the premises that are made up of these, too, are principles, just as the principles of composites are both matter and form, but so too are the bodies that are first composed out of the combination of these, that is, the elements.[235]

> **84b28** And just as there are some indemonstrable principles that this is that and that this belongs to that, so [there are indemonstrable principles] that this is not that and that this does not belong to that.

He says that just as there are affirmative indemonstrable premises, so too there are negative ones. And these [sc. negative indemonstrable premises] are the immediate ones. When he says 'this is that', he made the first word be [in reference] to that which is the subject for the predicate, as if we were to say 'the human being is an animal', and [when he said] that this belongs to that, [he made the first word be in reference] to the predicate, as if one were to say 'animal belongs to every human being', and likewise in the case of negative [premises].

> **84b30** So some principles will be that something is[236] and some will be that [something] is not something.

That is, some principles of demonstrations are affirmative and some are negative.

> **84b31** But when one must prove something, one must take what is predicated primitively [– let it be C –] of B, [and A[237]

similarly of it. For one who always proceeds in this way, no premise or [term] that belongs outside of A is taken in the proof, but the middle is always thickened until they [all] come to be indivisible and unitary.]

The task set for us is to prove through these [considerations] how the middle term is situated in regard to the extremes, in the three figures for both affirmative and negative syllogisms. And he says that the middle term never falls outside of the predicate in affirmative [syllogisms], but that in the first and second figure it always falls outside of the subject, while in the third [figure] it falls outside of neither the 268,1 predicate nor the subject. Now, having already made distinctions concerning the things said in affirmative [syllogisms], we will now speak in this way concerning the negative [syllogisms] too. In Aristotle, for affirmative [syllogisms, for a term] 'to fall outside' means [for it to be] of wider extent, and this is to be predicated [of something]. Now, he says, in an affirmative syllogism the middle will never have been predicated of the major [term]. For example, if one were to prove in the first figure that A belongs to B, we will take a middle term for both. And the way in which the major [term] A is predicated in a thesis or a conclusion is the only way in which it is predicated in the syllogism.[238] For A will have been predicated of C and C of B. And clearly, an affirmative [conclusion] is not proven in the second [figure], so the middle term will never have been predicated affirmatively of the major together with the minor [term] but of one or the other (whichever it happens to be).[239] But in the third [figure] the middle term will have been predicated neither of the major nor of the minor [term] since it must serve as subject for both. This is how things are for affirmative [syllogisms]. But, he says, in the case of negative [syllogisms], in the first figure, the middle term never falls outside of the predicate in that sense, but the manner in which the major [term] is predicated in the thesis or the conclusion, is how it always is in the premises too.[240] And in the third [figure], he says, it falls outside of neither the minor nor the major [term]. That is, it is of wider extent than, and is predicated of, neither of them, since in the third figure the middle term serves as subject for both [of them].[241] But, he says, in the second figure the middle term never falls outside of the minor term. Now the Philosopher said that in this case it is no longer possible to interpret 'to not fall outside' as 'to not be predicated of'. For in the second figure the middle term is predicated of both. But if it is predicated of the two, how is it that he says that it does not fall outside of the minor [term]? Now he said that we could not construct this in any other way than by understanding 'falling outside of the minor' differently here, namely as [being used] instead of 'the middle term never changes its position',[242] and [he says] that for the middle term to fall outside of the minor [term] is for the middle term to not switch position with the minor [term]. Now this, he says, does not occur in the second figure. For just as in the third [figure] the minor term is never a predicate but, just as in a thesis the minor term always serves as a subject,[243] similarly in the second figure syllogism the subject stays [where it is] and never switches its position. It is clear that this[244] is so in the first figure too. For in this the minor [term] is always a subject as well. Now these are the things that have been said here, and they do not extend further than what has been said in the [writings] concerning the three figures concerning the position of the three terms, and the [points] inserted here are not for the sake of anything that is required to be among the things useful in demonstration, or so it seems to me.

> **84b35** It is unitary when it is immediate and there is a premise that is one in the strict sense or is immediate. [And just as in other [contexts], the principle is something simple, though this is not the same thing everywhere ...]

Since he said that 'the middle is always thickened, until they come to be indivisible and unitary', in these lines he makes an additional point that in [the context of] a syllogism, a unit means an immediate premise. For when it takes additional middle terms, the syllogism is thickened down to the point at which you arrive at immediate premises, which are principles of the syllogism and are as it were certain indivisible unities.[245] For, he says, just as in number the principle and what is indivisible is the unit and in a line it is a point and in time it is the present and in something else it is something else, so too, in a syllogism the principle and what is indivisible are immediate premises. For these cannot be divided into two premises as mediated ones can.

> **84b38** ... but in weight it is a *mna* and in a song, the quarter tone [and something else in something else, so in a syllogism the unit is an immediate premise ...]

He says that the *mna* is a principle of weight in order to give an example. For in general there is no principle for these things, since magnitude is infinitely divisible, and if this is so, there is nothing that is first and is the principle of weight. But if there is [one], it is by virtue of our custom, as even now we employ an eighth of a carat as a last and indivisible weight. For we make a division up to this point. A quarter tone is the first perceptible sound which our hearing can discern. We say that this is a fourth of the whole tone.

> **85a1** … and in demonstration and scientific understanding, intellect. [So in syllogisms that prove that something belongs, nothing falls outside …]

For he says in the introductory chapters that 'not only do we say that there is such a thing as scientific understanding but also a principle of scientific understanding, by which we know the definitions'.[246] It would follow that the immediate premise is the principle of syllogism in general. For in dialectical [deductions] it is a reputable immediate premise. And of scientific understanding and demonstration the principle in the strict sense is intellect, by which we grasp the common notions. Why doesn't he also say that opinion is the principle of a dialectical syllogism? For just as intellect grasps the common notions on which demonstrations are based, so opinion [grasps] the dialectical and reputable premises. We say that he did not say that the immediate premise is the principle of dialectical syllogism but [that it is] of syllogism in general.[247] So the immediate premise is a principle insofar as it is [a principle] of syllogism in general.

> **85a3** … but in negative [syllogisms] nothing falls outside of that which must belong [for example, if A does not belong to B, through C (that is, if C belongs to every B, and A to no C); again, if one must [prove] that A belongs to no C, one must take a middle term for A and C, and it will always go on in this way.]

'That which must belong' refers to the predicate, for this is that which belongs to the subject. Now he says that in the first figure nothing falls outside of this, that is, nothing is denied of it. For in the first figure there is denied only the major of the middle term.[248]

> **85a7** But if one had to prove that D does not belong to E by means of C's belonging to every D and [its belonging] to no E or not to every E,[249] [it will never fall outside of E; rather it is this to which it must not[250] belong.]

He moved on to the second figure, which is why he changed the letters. And D is the major term, E the minor term, and the middle term is C.[251] He calls that to which it must not belong E. For the major term is denied of the subject in the theses and the conclusions. Now, he says, the middle term will never fall outside of the minor term, E. And we say what 'fall outside' means for him here, that [it means] to change the order.[252] For it falls outside of the major term. For although in the conclusion the major [term] is the predicate, in the premises he makes it the subject. However the minor [term] is kept as the subject in the premises.

85a10 But in the third way it will never fall outside of either that of which it must be denied or that which must be denied [of it].[253]

Here he again calls 'to be predicated of' 'to fall outside'. For it is not possible for the things said in these lines to be constructed in any other way. For in this figure the middle term always serves as subject, so that it is predicated of neither of the extremes.

Chapter 24

85a13 Since there is universal demonstration and particular demonstration, and also affirmative demonstration and privative demonstration, [there might be a dispute about which is better, ...] 271,1

Here he investigates three problems worthy of discussion, which bear on the discussion concerning demonstration. The first is which kind of proof, universal or particular, is superior and more suitable for scientific understanding,[254] the second is whether affirmative or negative [demonstration] is more suitable for scientific understanding,[255] and the third is whether direct proof or proof through an impossibility [is more suitable].[256] He goes through the discussion of these [matters] in great detail. He first considers [the issue of] the universal and the particular. He first presents a persuasive argument in regard to opposites, trying to develop the argument to the effect that partial proof is superior to universal proof, and is more suitable for demonstration. Then he utterly refutes any persuasiveness that these arguments have and proves that universal proof is superior. There are three ways in which he tries to develop the argument that particular [proof] is better. For if, he says, per se demonstration is better than one by virtue of something else and is <more> suitable for demonstration, and partial [demonstration] proves [something] per se, particular [demonstration] would be superior and more suitable to demonstration than universal [demonstration]. It is clear that per se [demonstration] is particular. For, he says, someone who proves that Koriskos is cultured proves per se to a greater extent than does one who proves that human being is cultured. For cultured does not belong to him insofar as he is a human being, but insofar as he is Koriskos. Similarly, we know per se that Koriskos is rational, but that a human being is rational we do not know per se but by virtue of something else. For we know that these particulars, for example, Socrates, Koriskos, Alcibiades, and the rest, are rational from themselves, however [we know] the universal, that every human being is rational, from the particulars. This is the first argument.[257] The second[258] is as follows. If, he says, universals, for

example, triangle in general, do not exist but only have their being in thought (for there is no triangle itself in itself[259] that is something different from the isosceles and equilateral and scalene, but there is also no isosceles besides the bronze isosceles or the wooden one or some such thing, and likewise in the other cases too, so the universal is among the things that do not exist, but particulars are existing things), and demonstration that is concerned with what exists is better than one that is concerned with what does not, it follows that particular [demonstration] is better than universal [demonstration].
272,1 The third puzzle is to the effect that that demonstration is better about which one cannot be mistaken, yet we are mistaken concerning universals, for demonstrations concerning universals produce accounts as if they exist in [their] subsistence, and it is a mistake to speak of what is not as though it is.[260] So in this respect too particular proofs are better than universal ones and are more appropriate for scientific understanding.

Now that he has laid out the puzzles in a persuasive manner he solves the puzzles as follows, beginning with the first. It is false to say that what is proven belongs to the particulars by virtue of themselves, but to the universals not through themselves but through the particulars. For the isosceles has the three angles equal to two right angles not insofar as it is isosceles (for if this were so it would belong to nothing else) but, insofar as it is isosceles, it will have equal [angles] at the base. However it will have the three angles equal to two right angles not insofar as it is isosceles but insofar as it is a triangle.[261] Similarly, being rational belongs to Socrates not insofar as he is Socrates (for if this were so it would belong to no one else) but insofar as Socrates is a human being. And for Koriskos to be cultured belongs per se to Koriskos, but this will no longer belong to the universal as well, for cultured does not belong to human being insofar as one is a human being. For this too is the reason behind the mistake: it is not possible to say that human being is cultured. For if you are talking about a particular human being, you are no longer referring to the universal but to either Socrates or Koriskos or someone else. So, he says, if an attribute does not belong to a universal, for example, for a human being to be cultured or for a triangle to be three cubits tall or some such thing, and someone then proves that such a thing belongs to the universal, such a thing is not a demonstration but a fallacy, since it says that there belongs to all what belongs to a particular. But in general if there is an attribute that belongs to a universal, such as rational to human being or having the angles equal to two right angles to triangle, it is not because it belongs to particulars that it belongs to the universal as well, but it is because it belongs to the universal that it belongs to the particulars too, if it indeed belongs insofar as it is a triangle, not insofar as it is isosceles or scalene.[262] For when this is eliminated the attribute too would be eliminated. But as things are this is not eliminated. Rather, it is eliminated when triangle is eliminated.

This is how he refutes the [argument offered by] the first puzzle. [He refutes] the second by saying that it is false to say that the universal is among the things that do not exist. For if, he says, a 273,1
universal were to be an equivocal term, as 'Ajax' is applied to both the son of Oileus and to the son of Telamon and 'dog' is applied to the terrestrial dog, the sea-dog, and the stellar dog, the universal would be among the things that do not exist in reality. But if there is a single definition of universal and there is a single common nature that inheres in many things, it is false to say that the universal is among the things that do not exist. Rather, it is a different and common substance,[263] but it is neither separable nor subsisting in itself; rather it runs through all particulars and subsists in them. And just as we say that the categories of accident (quantity, quality, and the rest), are certain natures, each distinguished from other things, yet each subsisting not in itself, but by having its subsistence in the particulars – and we do not say that it [the nature] exists in no way beyond its not subsisting in itself, but we say that each is among existing things and subsists by virtue of being defined by a common account, even if it has its subsistence in the particulars – so we say that even in regard to the universals there is a nature of triangle that is different from and apart from (*para*) all particulars, I mean, from isosceles, scalene, and equilateral, but not having its subsistence outside of them.[264] For it is possible to isolate it from all of the particulars by means of a definitional account. Therefore, the universal is not among the things that do not exist, but rather <exists> to a higher degree than particulars do, given that it is indeed imperishable, and all particulars are perishable, and that which is imperishable is among things that are to a higher degree than that which is perishable. Yet, if the universal is among the things that are, the cause of the mistake is not the proof of the universal which proves [something] about what does not exist as though it does. But if when others prove the universal, they produce their account about it as though [the universal] were something subsisting itself in itself apart from the particulars, this very fault lies not in the demonstration but in the one who interprets it.[265] For the demonstration does not posit the universal as separable. If someone interprets it as concerning something separable, he himself is the cause of his mistake.

85a15 And things are also this way in the case of the [demonstration] which is said to demonstrate [and that which leads to an impossibility. So let us first investigate the universal and the particular. Once we clarify this, let us also speak about that which is said to prove and that which [leads] to an impossibility.

Perhaps it might seem to some who look into things in the following way that particular demonstration is better. For if the better demonstration is that demonstration by which we have scientific understanding to a higher degree …]

['The demonstration that is said to demonstrate'] is instead of '[in regard to] direct [demonstration]'. For demonstration that proves the fact directly is also demonstration in the strict sense. For reduction to an impossibility does not prove [the conclusion] itself, but eliminates the opposite.

30 **85a22** … for this is the virtue of demonstration. [And for each thing we have a higher degree of scientific understanding when we know it per se than when we know it by virtue of something else. For example [we know] cultured Koriskos [to a higher degree] when [we know] that Koriskos is cultured than when [we know] that a human being[266] is cultured, and similarly in other cases.]

This, he says, is the virtue of demonstration, when it[267] comes about not through something else, but through itself.

274,1 **85a26** And universal [demonstration] shows that something else, not the thing itself, happens to be [something or other], for example, that the isosceles [is such and such] not because it is isosceles but because it is a triangle, while particular [demonstration] [shows] that [such and such is the case] because it is itself. [So if that which is per se is better, and particular [demonstration] is that sort of thing to a greater extent than universal [demonstration], particular demonstration would be better.]

Universal proof, he says, [comes about] not by virtue of itself but
5 because of particular proof. For when someone proves the things that belong per se to isosceles through the universal, for example, that an isosceles triangle has equal angles at the base, he proves it insofar as it is a triangle, not insofar as it is isosceles. Clearly, he employs a false example, since it was not possible for him to employ a true one. For he did not prove the things that belong to isosceles[268] as though they apply to triangle, but, as he next says, when someone proves
10 that that which does not belong to the universal does belong, such is not even a demonstration. For nothing [that belongs] by virtue of something else is proven to belong to the universal.

85a31 Further, if the universal is nothing apart from the individuals, and demonstration instils the opinion that that by

> which it demonstrates is something, and that there is a certain kind of nature that belongs among the things that are of that kind[269] [for example, [it proves something] of a triangle apart from the particulars and of a figure apart from the particulars, and of number, apart from the particular numbers, and [demonstration] concerning what is is better than that concerning what is not, and [demonstration] through which we will not be deceived is better than that through which we will, …]

He intertwined the second and third puzzles, namely, that the universal is nothing apart from the particulars, and that the cause of the mistake is our positing as existing what does not exist. For we posit the universal as something existing in itself, apart from the particulars.

> **85a36** … but universal [demonstration] is this sort of thing.

'This sort of thing' – [i.e.] the sort of thing he talked about [– a demonstration] that posits as existing what does not exist and for this very reason is the cause of our mistake. In each case he independently develops the argument of how this happens, and he first posits the universal as not existing by inductive [arguments].

> **85a37** For, as in the case of proportion, they proceed to prove, for example, that whatever is this sort of thing [which is neither line nor number nor solid nor plane, but something apart from these] will be proportional. [So if this is universal to a higher degree [than the particular demonstration], but concerns what exists to a lesser degree than the particular one does, and if it instils a false opinion, it follows that universal [demonstration] will be inferior to the particular demonstration.]

Just as, he says, the geometrician proves that, if four lines make a proportion, they will also make a proportion when alternated, this would also be the case if there were four planes or four solids, this point generally holds in regard to universal items[270] because if the four are a certain kind of thing, for example, quantity or magnitude, 275,1
and, in addition, neither quantity in general nor magnitude exists itself in itself outside of line or plane or solid, and yet they develop the argument concerning them as though it were something that exists apart from these things, so, what happens to them [when demonstrating about] proportion is the very same thing that happens to all of those who demonstrate something universal. For when they assume that horse and human being and dog and the rest are an animal they also take animal in general, which they posit as different from the things mentioned, to be existent even though it is not different from them, so that they posit <as existing> what does not

exist. Those who ascend from individuals to Forms do likewise. So if universals have to do with what does not exist, and particulars with what exists, and what exists is superior to what does not exist, particular demonstrations are superior to universal ones. He says that it is for this reason that the universal is something that instils false opinion, since it requires that we posit what does not exist as something that exists. So it too is the cause of our mistake. So for this reason too particular [demonstration] is superior.

> **85b4** It is rather that, first, the other argument is no more [applicable] to the universal than to the particular.[271] [For if [being equal] to two right angles belongs [to a triangle] not insofar as it is isosceles but insofar as it is a triangle, the one who knows this about isosceles knows it, as such, to a lesser degree than the one who knows it about triangle.]

At this point he still needs [to give] the solutions to the puzzles. He devotes some time to the first [puzzle], which states that partial [demonstration] demonstrates per se while universal [demonstration] demonstrates by virtue of something else, for [it demonstrates] by means of the partial [demonstration]. Since he distinguished two arguments (for, as we already said,[272] that which we divided into two he put forward in an intertwined manner) he said that 'the other argument', that is, the first, which we now mentioned, is no more [applicable] to the particular than to the universal. He [however] states the converse. For, while he should speak in the way we said [i.e. 'no more to the particular than to the universal'], he [instead] says 'no more to the universal than to the particular'.[273] For he wants to prove that the universal demonstration is per se, not by virtue of something else. What does he mean when he says 'no more [applicable] to the particular than to the universal'? To be sure, he will proceed to show that insofar as it is a triangle, its three angles are equal to two right angles per se, but not insofar as it is isosceles. Rather it belongs to isosceles by virtue of something else, and this is triangle. How is it that he here says 'no more applicable to the particular than to the universal' since both of them are per se? Now
276,1 I say that the discussion bears on his example. For since in the example – I am referring to 'Koriskos is cultured' and 'human being is cultured' – 'the particular has the [attribute of being] cultured per se, but the universal has it through the particular' – for this reason he says that in the case of such examples, in which the attribute does not belong without qualification to the universal, the particulars will not have the [characteristic of being] per se to a higher degree but in the cases in which it belongs in general to the universals too, it belongs because of them to a higher degree than it does because of the particulars. For when something belongs to both a universal and

a particular, it does not belong to the universal because of the particular; rather, on the contrary, [it belongs] to the particular because of the universal.

> **85b7** And in general, if [triangle] is not [something or other] insofar as it is triangle, yet he proves that it is, that would not be a demonstration.

Since prior to this he had agreed that there is what is per se in the particular, too, and he said that it does not exist to a higher degree than in the universal, he now eliminates this [view], that the particular [substance][274] is per se at all, by means of what is called 'objection and counter-objection', and having first dealt with the counter-objection, he now [deals with] the objection. For, he says, I am saying that the per se does not exist at all in the particulars, for in the example put forward, being cultured does not exist at all in human being but only in the particular, so to say that it belongs to the universal by virtue of something else is false, since it does not belong to the universal at all. For if there is something that belongs to the universal at all, by all means it will belong per se too, and to the particular because of the universal, as rational does to human being and perceptive to animal and equal to two right [angles] to triangle, and likewise in all cases. The exegesis of the text is as follows. 'And in general', he says, 'if [triangle] is not [something or other] insofar as it is triangle', yet he proves this, 'that would not be a demonstration'. That is [to say], if the attribute does not belong to the universal insofar as it is such, for example, say, having the perimeter, say, of five cubits, to triangle insofar as it is triangle, or being cultured to human being insofar as he is a human being, and consequently someone talks about what does not belong as though it belongs, such a person would not be demonstrating. For that sort [of thing] is false, and just as that which is true is never refuted, so the false is never demonstrated. And that which does not belong to something at all will not belong to it by virtue of something else.

> **85b8** But if it is, the one who knows each thing, insofar as each one belongs, knows it to a higher degree. 277,1

That is, if when there is an attribute that is in the universal insofar as it [the universal] is a certain kind of thing, someone demonstrates that, for example, having two sides greater than the remaining one, of triangle, and someone who knows something per se knows it to a higher degree, then one who knows the universal knows to a higher degree. And that by which we know to a higher degree is superior. Therefore universal demonstration is superior.

85b9 So if triangle is of wider extent and there is the same definition and it is not a triangle equivocally etc. [and two [right angles] belongs to every triangle, triangle would not have such angles insofar as it is isosceles, but isosceles would [have them] insofar as it is a triangle. So the one who knows universally knows [something] to a higher degree insofar as it belongs, than the one [who knows] by virtue of a particular. Therefore universal [demonstration] is better than particular [demonstration].]

He takes up the argument and draws the conclusion. So, he says, if triangle is of wider extent than isosceles (for it belongs to scalene and equilateral too) and this common species of triangle is not an equivocal term but there is one definition and one nature, and having the three angles equal to two right angles belongs to every triangle, then clearly such a thing does not belong to triangle because of isosceles (for then it would not belong to every triangle but only to isosceles), but [it belongs] to isosceles because of triangle. So for this reason universal [demonstration] is per se to a higher degree and is better than particular [demonstration]. He was correct to add 'not ... equivocally'. For being a substance belongs to every dog, but not insofar as it is a dog (for there is not a particular substance of a dog in general which is called a dog,[275] as there is of animal, but it is an equivocal term), but because being a substance belongs to this kind of animal and to that kind of animal. However, 'triangle' here is not equivocal, but there is one particular common definition of it.

85b15 Another point is that if there would be a single definition and the universal is not a [matter of] equivocity, it would not exist any less than some of the particulars, but even more, insofar as imperishables are among the former, but it is rather the particulars that are perishable.

This is the response to the second puzzle, which stated that the universal is something that does not exist. So, he says, if the universal is not an equivocal term like 'Ajax' or 'dog' or some other such thing, but there is a single common definition of the universal, he says that not only will the universal not be among the things that do not exist but it would exist to an even higher degree than what is particular, since particulars are perishable, while universals are imperishable.

278,1 **85b18** Further, there is no need to assume that this is something apart from those [particulars], on the grounds that it refers to one thing, [any more than there is in the case of the other items that signify not the what but a quality or a relation or an action.]

He says that, because the puzzle asserts that those who produce arguments concerning the universal posit it as something that itself belongs per se and as something different from the particulars, nonetheless even if it is something different from the particulars and there is a single determinate nature different from the particulars, they do not posit that it is separate and subsists by itself outside of the particulars, but that it is different in both substance and in definition – nonetheless, [he says], it does not differ from the particulars in its subject, just as we say that accidents, too, are different from their substance but are not separate from it. So those who posit [it] do not posit a falsehood, and the cause of the mistake is not within us insofar as we posit the universal, which is something that does not exist, as existing.

> **85b21** But if this [conclusion is drawn], the demonstration is not the cause; rather the one who interprets it is.

That is, if one were to take it as subsistent separable from particulars, the cause of taking the universal to be separable would not be the demonstration but the one who incorrectly interpreted the demonstration. It is just as if someone were to discuss quality as though it were a certain nature that both produced and underwent many things, and then one were to interpret this and suppose that the one who was talking about quality posited it as something subsisting by itself. The cause of such a supposition and false opinion would not have been the one who spoke but the one who incorrectly interpreted it. Likewise if[276] demonstration produces an account in regard to universals and someone else thinks that the universal is separable, he is the cause of his own falsehood. However, the demonstration does not posit falsehood. Yet, one might with good reason investigate how the things said here square with those [said] in the introductory chapter of the treatise *De Anima*. For in these [lines] he says that the universal <not> only exists and is a substance[277] different from the particulars in its definition, but that it is imperishable and is primary, in regard to the particulars, as he says in what follows, while in *De Anima* he says totally the opposite, that either there is no universal at all or, even if there is, it is posterior to the particulars, and for this reason it is obvious that it would not be imperishable, even if particulars are perishable. The text in *De Anima* is as follows: 'One must be careful to take note of whether there is a single definition for it, as there is for animal, or a different one for each, such as for horse, dog, human being, god, the universal animal being either nothing or posterior. This is how it is if some other common 279,1
item is predicated, as well.'[278] How again is it possible for a universal to exist, unless it separated from the particulars but has its subsistence in them? Again, how would it be possible for it to be imperish-

able except in the definitional account? For what is individually qualified is not, so described, imperishable in number and in this way too the individual form does not persist as imperishable, but when the subject perishes it is entirely necessary that it [i.e. the form] too perishes since it has its being in it [i.e. the subject]. That which comes to be is not the same in number but it is the same sort of thing. A boat [offers] an example of what is individually qualified, since it is changed in its timber, and in the end is totally changed but preserves the form it had from the start. For in this case the form of the boat is preserved as the same in definitional account, but it is not [the same] in number. Now we say that this is how it is in the case of natural forms as well. For if they are not separable but have their subsistence in particulars, when the subjects perish, it is entirely necessary that they too perish, but [there is] imperishability in respect to the definitional account alone, which always remains the same. For just as the whiteness that now inheres in common in all the white bodies is not the same in number as that in the times of Plato, so it is not the case that the animal now is [the same as] the [animal] then, nor is this so for anything else among such things. Now either they are imperishable and do not exist in particulars, or they are in them and are not imperishables. These [matters] will be more thoroughly and completely examined by us elsewhere.[279]

Let us go on to the next [part] of the account.

> **85b23** Further, if a demonstration is a probative syllogism of the cause and the reason why, and the universal is the more important cause etc.

Having solved the puzzles by which particular demonstration seemed to be superior to universal [demonstration] he now intends to establish the very theorem per se that universal [demonstration] is superior to particular [demonstration]. He proves this through a number of arguments, of which the first is as follows. He says that if demonstration is nothing other than a syllogism that establishes the cause of the fact, and the universal is the primary cause of the fact, and we especially know facts when we learn the primary and most important cause of their being the case, universal [demonstration] is better and more appropriate to scientific understanding than particular [demonstration]. That the universal is a more important cause he proves from [the fact] that the universal belongs per se to a higher degree, as was mentioned above. For if universal proof is per se, and what is per se is a cause to a higher degree than what is not per se, it follows that universal [demonstration shows] the causes to a higher degree than does particular demonstration. And he also proves in a different way that the universal is primary and a more important cause. For when we investigate the reason why, we ascend

this far in investigating the reason why: until we reach the first 280,1
cause. And when we hear this we cease the rest of the inquiry since this is the most important cause. For example, why did so and so go out? In order that he might enter the agora. Why did he enter the agora? Perhaps in order to buy vegetables.[280] And why do this? So that when he is eating he might stimulate the juices in his belly. Why stimulate the juices in his belly? So that he might be healthy. And if we are considering only the good of the body, when we hear this, we have the cause that is the most aimed at and no longer seek another. If our eye is on the good of the soul, we say, why does he want to be healthy? In order that he might be able to be active in accordance with virtue. This, so that he might be happy.[281] And this, happiness, is the most primary and most important cause of entering the agora, and when we hear this we have the cause of the fact. And this is how it is not only in regard to the final cause but to all of the others as well. For example, how is it that the statue came to be? Because it is from bronze. And how did bronze produce the statue? By being poured. Why was it poured? Because it is metal, for certain metals, including bronze, are pourable. And, again, when we hear this we have the most important material cause. This is how it is for the efficient cause too. Why was a human being generated? Because he was conceived and moulded in the mother. And why was he conceived? Because the male came together with the female. Why did this happen? Because the male has the power of emitting seed, and the other has the power of receiving it. And when we hear this we have the most important efficient cause. This is how it is for the formal cause too. Why is Socrates a living, perceptive substance? Because he is a human being, and a human being is that sort of thing. Why is a human being that sort of thing? Because [a human being] is also an animal. Why is an animal that sort of thing? Because that is the essence and the definition of an animal. So if for all of the causes we especially know when there no longer belongs some other cause of the fact, we call the very last one especially and primarily the more important cause, and this is a per se cause, because such a thing is not by virtue of something else but because of itself. Universals are such things.[282] For <in> the demonstrations of causes, these are the last items at which we arrive. For example why does this bronze isosceles have the exterior angles equal to four right [angles]? Because every triangle does. So does the triangle have this attribute per se? Not at all, but insofar as it is a rectilinear figure. This is no longer because of something else but because of itself. So this is a cause to a higher degree, and this is universal. Therefore this [i.e. the universal] is the cause to a higher degree. That which is the cause to a higher degree is better than that which is the cause to a lesser degree. Therefore the universal is better than the particular.

281,1 **85b23** … of the cause and the reason why

'Of the cause and the reason why' is a pleonasm.

85b24 … the universal is the more important cause.

The next and last thing is to infer 'Therefore demonstration of the universal too is better, for it is of the cause, to a higher degree. But before he infers the conclusion, he first shows in what way the universal is the more important cause.

> **85b24** For that to which something belongs per se is itself its cause. But the universal is primary. Therefore the universal is the cause. [So [universal] demonstration is better. For it is of the cause and the reason why to a higher degree.]

If the attributes that belong to the universal belong per se, and that to which something belongs per se has no other cause of what belongs to it, but is itself its own cause,[283] and this is primary, and a primary cause is the cause to a higher degree, it follows that the universal is a cause to a higher degree. That what belongs to the universal belongs per se, but what [belongs] to the particular [belongs] because of the universal, has been often stated.

> **85b27** Further, we investigate the reason why up to this point, and we then think that we know when it is not the case that this is because something other than it either occurs or is.

He proves by means of these [considerations] as well that the universal is a more important cause. For, he says, when we investigate the reason why we ascend up to the universal, and when we reach it, we stop our inquiry since we have found the most important cause of the fact. So the universal is a cause to a higher degree. 'When it is not the case that this is, because something other than it … .' That is, when it is not possible to assign some other more universal cause, but that which has been given is the last and most important cause. For example, triangle is the cause of the triangle's having the three angles equal to two right [angles], and when we have ascended to this point we find no cause beyond it. Similarly rectilinear figure [is the cause] of the exterior [angle]s' being equal to four right [angles] and we seek no cause beyond it. He says 'either occurs or is' since of causes some belong as beings, not as occurrences, as in both mathe-
282,1 matics and physics (for triangle and figure are in what is rendered as the causes of why the three angles are equal to two right [angles]) but some are not but occur, as in things that are done and in the example we already gave, that he went out in order to enter the agora, and

this, in order to buy vegetables, in order to get the juices going in the belly, in order to be healthy, in order to practice virtue, in order to be happy. These causes *are* not, but they *occur*.

> **85b29** … for in this way that which is last is in this way an end and a limit. [For the sake of what did he come? In order to get money, and this was to give back what he owed, and this to avoid acting unjustly. And if we go like that, when [he] no longer [came] because of something else nor for the sake of something else, we say that he came, and that [something] is the case, and occurs because of this as an end, and that it is then that we especially know why he came.]

He says 'in this way that which is last in this way', that is, that which is the last in the accounts of the causes, 'is an end and a limit' of the causes, for which there is no more important cause. The example, which he thereupon gives, is of the final cause.

> **85b35** If this is how things are for all of the causes and reasons why, and in the case of that for the sake of which we especially know [when it is] in this manner, then in the case of the other [causes] too we especially know when this no longer belongs because something else [does].[284] [Now when we know that the exterior [angles] are equal to four [right angles] because it is isosceles, there remains the reason why the isosceles [has this attribute]: because it is a triangle, and this [has this attribute] because it is a rectilinear figure. If this is no longer because of something else, we especially have knowledge. And we do so universally. Therefore universal [demonstration] is better.]

Since, as I said, he produced an example of the final cause, he wants to deduce on that basis that this is how things are for all of the other causes too. He says that if [the situation] is like this in the case of all of those causes that we are wont to render when we are asked the reason why – that is, if this is how we render [them] for all of the causes, when we are asked, and we showed that for the case of the final [cause], we especially know when we know the last cause, after which it is not possible to render another, clearly this is how things are for the others. We ourselves gave examples of all of them. 'When this no longer belongs, because something else [does]', that is, when that which is said to belong no longer belongs because there is something else, for example when having the three angles equal to two right angles no longer belongs because of anything else but because of triangle. And similarly the exterior angles being equal to four right angles [belongs] not because it is an isosceles, nor because it is a triangle, but because it is a rectilinear figure, but no longer because of something else. Such items are an account of the formal cause, for having exterior angles equal to four right angles is constitutive of the species[285] of rectilinear figure, as having the interior [angles] equal to two right [angles] is of triangle. So if when we know 283,1 the last cause, after which there is no other, we then especially know, and if the universal is this sort of thing (for figure or triangle is universal), we especially know when we know the universal, and that [sort of] demonstration is better.

86a3 Further, the more it is particular, the more it falls into indefinites. But the universal [falls] into what is simple and a limit. [And insofar as things are indefinite, they are not objects of scientific understanding, but insofar as they are limited, they are objects of scientific understanding. Therefore, insofar as they are universal, they are objects of scientific understanding to a higher degree than they would be insofar as they are particular.]

Further, with these remarks too he shows that universal demonstration is superior to the particular one. For a universal demonstration, he says, proceeds to a limit and unity, but a particular [one proceeds] to indefinite things, for individuals are indefinite. But that which is indefinite is unknowable by scientific understanding.[286] Rather, scientific understanding is of things that are limited and determinate. So if [demonstration] that proceeds to the limit and unity is superior, and that which proceeds to the indefinite is inferior, and to the extent to which something is more universal, to that extent it is closer to the limit and unity, and to the extent to which it is more particular, it is close to the indefinite and the many, clearly, then, from these considerations too it follows that universal demonstration is superior to particular [demonstration].

86a7 Therefore universals are more demonstrable. And the greater the degree to which it is of demonstrables, <the greater the degree to which> it is a demonstration.

They are more demonstrable, since there is no demonstration of indefinites. So those that are closer to the unity are those that are more demonstrable. And those that are closer to the indefinites are those that are less demonstrable. But universals are closer to the unity. Therefore universals are more demonstrable.

86a9 ... for relatives are 'to a higher degree' correlatively.

Since he said that there is a demonstration to a higher degree of things that are demonstrable to a higher degree, he establishes the same point. For if relatives are by nature correlative, then clearly as the one is, so is the other, too. For if this one is more of a friend than that one, the one to whom he is more of a friend is also more of a friend to him. So if demonstrables are demonstrable by demonstration, there will be a demonstration to a higher degree of the things that are demonstrable to a higher degree.

> **86a10** Again, if the demonstration by which we know one thing and another is to be preferred over that by which we know only the one thing, and the one who has the universal demonstration also knows the particular etc. [while the other does not know the universal, it follows that the first would be the one to be preferred in this sense, too.]

This is another argument. He says that the demonstration by which
we know more things is superior to that by which we know fewer. For 284,1
example, the one who knows about all human beings is superior to the one who knows about only one or several. So we know the particulars too by means of universal demonstration, whereas someone who knows a particular does not know the universals. So universal [demonstration] is superior to particular [demonstration] in this sense too.

> **86a13** Another point is this. To prove more universally is to prove by means of a middle term that is closer to the principle. [What is immediate is closest, and that is the principle. So if [the demonstration] on the basis of a principle [is more precise] than that which is not on the basis of a principle, the demonstration which is more on the basis of a principle is more precise than that which is less [on the basis of a principle]. Universal demonstration is that sort of thing to a higher degree. Therefore universal demonstration is superior. For example, suppose one had to demonstrate A of D. B and C are middle [terms]. B is higher, so the demonstration through it is more universal.]

This is another argument. If every demonstration is on the basis of principles, and of the principles, the first principles are the most important (for the one who knows them knows the ones under them too, for as we said, the one who knows the universal also knows the particular, but the one who knows the latter does not necessarily know the universal) the demonstration that is on the basis of the first principles and causes is the most important. Demonstrations through the universals have their middle terms closer to the first principles. So if those that are on the basis of the first principles are especially demonstrations, clearly those that are on the basis of those that are closest to the first [principles] are demonstrations to a higher degree than those that are not on the basis of them. To the extent that the demonstrations are more universal, to that extent the middle terms are closer to the first principles. To the extent that the middles are more particular, to that extent they are farther. For example, suppose we wanted to prove that a human being is a body. There are several middle items by means of which we can prove this. For it is proven through living, say, and through animal and through rational. But animal is closer to the principle, that is, to the essence, than is rational, and living [is closer] than animal. So if the demonstration that is through what is closer to the principle is more important and better, and that is more universal, the universal [demonstration] will be better. For he says that it is more universal because it employs the more universal middle [term]. It must also be pointed out that he here uses 'more universal' and 'more particular' in a different sense. For he does not call 'particular' that which has a particular determination such as 'some …' and universal that which has [the determination] 'all' but instead that which employs either more universal middle terms, as we already said, or even more universal extremes, for example, 'every human being is an animal' would be more particular than 'every human being is a substance'.[287] What is under investigation is whether the deduction that proves the particular is different from that which proves the universal, for example, 'every human being is an animal' and 'some human being is an animal'.

285,1 **86a22** But some of the things that we have stated are formal. [It is most clear that universal demonstration [is a demonstration] in the more proper sense on the grounds that when we have one of the premises that is prior we in a sense know and have the posterior too, potentially. For example if someone knows that every triangle [has interior angles equal] to two right angles, there is a sense in which he knows potentially that isosceles too [has interior angles equal] to two right angles, even if he does not know that the isosceles is a triangle. But the one who has possession of that [latter] premise does not in any sense know the universal, neither potentially nor in actuality. Also, the universal is intelligible, but the particular has its end in perception.]

He says that some of the arguments that he stated are formal, not because they are persuasive though not true, rather, [because] they are true though too general and apply not only to demonstrables but also to certain other [matters].[288] For example, [the premise] 'that by which we know something and something else' is superior 'to that by which we know it alone'. For this would apply to [matters] that are not demonstrable, too. For example, knowing carpentry and stonecutting is superior to knowing one or the other of these alone, and likewise in very many other cases, and these [matters] are not demonstrable. But it would apply to demonstration too, since the one who knows the universal knows more than the one who knows the particular. Likewise: 'Further, we investigate the reason why up to this point, and we think that we know when it is not the case because something other than it either occurs or is.' For this also applies to [matters] that are not demonstrable, as when it is one of the individuals. For example, why did so and so go out? In order to buy vegetables, etc. Such [matters], he says, are not proven in the strictest sense. Therefore universal [demonstration] is superior to [demonstration] that is particular for the following reason too – because the one who has the universal has the particular too, potentially, but the one who has the particular has the universal neither potentially nor in actuality, and even if it often happens that the one who knows the universal, and for that reason comprehends the particular, is unaware of some of the particulars through not having considered them – for example (one we have mentioned earlier), someone who knows that no mule conceives, but having found a mule that had its belly swell up will think perhaps that it had conceived, and the one who knows that the angles at the base of isosceles [triangles] are equal, but who found an isosceles [triangle] will think, as I said, through not having considered [the applicability of what he knows] that it does not have equal [angles] at the base. Similarly too, [for example], it is to a higher degree that the universal [demonstration] travels towards the intelligible and unitary. But the one regarding particulars [travels over] perceptible and indefinite [matters]. On these grounds, he says, the [demonstration] of the universal is especially proven to be superior to that of the particular.

Chapter 25

> **86a31** [Let this be as much as we say about how universal demonstration is better than particular demonstration.] But it is clear from the following that probative [demonstration is superior to] privative [demonstration]. For, other things being equal, let the better demonstration be …

Having completed [his treatment of] the first problem, he now moves on to the second, that is, that affirmative [demonstration] is superior to negative [demonstration]. He proves it as follows. He says that if there are two demonstrations, and one draws its conclusion through more middle terms and the other through fewer middle terms, other things being equal, that which is through fewer middle terms is superior to that which is through more. He added 'other things being equal', which is to say, if the middle terms in both demonstrations are similarly known, for if the [demonstration] that is through more middle terms has middle terms that are better known than those of that which is through fewer [middle terms], the [demonstration] through [the middle terms] that are better known is superior. But if they are similarly known, the demonstration through fewer middle terms is to be preferred, for it is this that is closer to the activity of intellect. For if intellectual activity is superior to deductive activity, and intellectual activity apprehends intelligibles in an immediate way, clearly the deductive activity through fewer middle terms is the one that is superior since it is closer to the immediate knowledge. For example, if we want to prove that A belongs to E, and here it is proven through the middle terms B, C, and D, and here through fewer [terms], F, G, that which is through F and G will be superior to that through B, C, and D, insofar as it is through fewer middle terms. That it is superior is clear from the following [consideration]. Prior middle terms are always better known. For example, if A, B, C, D, and E are successive terms, and A is predicated of all of them successively, that which is closer to A is better known. For that A belongs to B is better known than [that it belongs] to those that succeed, and, again, that A belongs to C [is better known] than that [it belongs] to D and E, and again, that A belongs to D [is better known] than that A belongs to E. Let the terms be living, animal, rational, human being. So if demonstrations through primary middle terms are better known, and those that are through what are better known are superior, it follows that the demonstration that proves that A belongs to D through <B and> C is superior to that [which proves] that A belongs to E through B, C and D. So as A and D are, through B and C, so A and E are through F and G, for the middle terms are equal in number. But the [demonstration that] AD is superior to the [demonstration that] AE, and the [demonstration] that AE through F and G will then be superior to the [demonstration] through B, C, <and D>.[289] So if the [demonstration] through fewer middle terms is superior to that which is through more, then, if it is proven that the negative [demonstration] is through more [middle terms] than the affirmative, negative [demonstration] is inferior to the affirmative [demonstration]. By 'more' I mean not more in number but in kind. For example, if there were a negative deduction that A [belongs] to no B, but B [belongs] to every C, and if it would be necessary to establish each of the premises, the affirmative [premise] will have been proven by affirmative middle terms, and the negative [premise] by a negative and an affirmative [premise]. Therefore the negative [demonstration] is proven through more [middle terms] – [i.e. more] in kind – than is the affirmative demonstration. So affirmative demonstration is superior to negative [demonstration]. This is also [proven] in a different manner. The [demonstration] that is independent and in need of nothing else for its perfection is superior

to that which is in need of something else. Now, the affirmative [demonstration] is in need of nothing else but is itself proven through itself. But the negative is in need of the affirmative. For a negation 287,1
is never proven without an affirmation, for nothing is proven on the basis of negations alone. So for this reason too affirmative [demonstration] is superior to negative [demonstration].

86a34 ... the one that is from fewer postulates or hypotheses or premises. [For if they are similarly known, knowing will occur more quickly through these, and this is to be preferred.]

That is to say, whether the premises are postulates or hypotheses or something else, the [demonstration] that is from fewer premises is superior without qualification to that which is from more, other things being equal, that is, when [premises] are all[290] known in the same way – either all reputable or all demonstrable or are taken in a similar way from what holds accidentally or from [what holds] per se, and everything else that belongs to premises. When these are the same, the [demonstration] that is on the basis of [fewer] middle terms and fewer premises is superior to that which is through more.

86a36 The universal argument for the premise that [the deduction] is better that is on the basis of fewer [premises] is as follows. [For if the middle terms are similarly known ...]

'Argument': in other words, reason and demonstration of [the fact] that the superior [demonstrations] are those that are on the basis of fewer [premises], other things being equal.

86a38 ... and those that are prior are better known, [let there be a demonstration that A belongs to E through the middle terms B, C, and D, and one that A [belongs] to E through F and G.]

For should we want to prove that living belongs to human being, with animal and rational as intermediates, it is better known that animal belongs to rational than [that it belongs] to human being. For [the fact] that animal belongs to human being is proven through [the fact] that animal belongs to rational. For that through which [an item is known] is better known than that which is known through that item.

86b2 A belongs to D in a manner similar to how A belongs to E. [But that A [belongs] to D is prior and better known than that A [belongs] to E. For the latter is demonstrated by means of the former, and that by means of which [something is demonstrated] is worthy of conviction to a higher degree. Therefore the demonstration through fewer [middle terms] is better, other things being the same.]

'In a manner similar': that is, through an equal number of middle terms. For, just as A has been proven of D through B and C, so A [has been proven] of E through F and G. And it has been assumed that the middle terms are similarly known. So if it is similarly known that A [belongs] to E through F and G, and that A belongs to D through B and C, and it is better known that A belongs to D through B and C than that A [belongs to] E through B, C, and D, it is therefore also the case that it is better known that A [belongs] to E through F and G than that A [belongs] to E through B, C, and D. For that [belongs] through fewer [middle terms], F and G. And that [which is known] through fewer [middle terms] is better known.

288,1 **86b7** Now they are both proven through three terms and two premises, but the one [demonstration] assumes that something is [the case] and the other both that something is [the case] and that something is not. [Therefore it is proven through more, so it is inferior.]

Having proven universally that, other things being equal, the [demonstration] through fewer [premises] is better – fewer in general, not specifying whether they are fewer in number or in kind – he now applies the argument to the very [matter] at hand, on the grounds that the negative [demonstration] is proven through both affirmative and negative [premises], but the affirmative [demonstration] is proven through affirmative [premises] alone. So if negative [demonstration] is [proven] through more, affirmative [demonstration] is superior.

86b10 Further, since it has been proven that it is impossible for there to be a syllogism when both premises are privative etc. [but there must be one of that kind and another that it belongs ...]

He does not present this as a different argument, besides that which precedes it, but as establishing that the negative [demonstration] requires the affirmative [demonstration] while the affirmative [demonstration] does not require the negative, on the grounds that it has been proven that a syllogism never arises from negative premises alone. The reason why having said 'further, since it has been proven' there is no apodosis for the 'since',[291] is that 'since' is said in the way that identifies the cause of what precedes it.

86b12 And again, in addition, this must be assumed: as the demonstration is increased there must be more affirmative

> [premises] etc. [but in every syllogism there cannot be more than one privative [premise.] For let A belong to no Bs, and let B belong to every C. So if you need to again increase both premises, you must insert a middle term. Let D be [the middle term] for A and B, and E for B and C. So clearly E is affirmative, and D is affirmative of B, but is privative in respect to A. For D [is predicated] of all B, but A must belong to none of the Ds. So A is D is a single privative premise. This is how things are for the other syllogisms too. For the middle term for affirmative terms is always affirmative for both, but for the negative term, [the middle term] should be negative for one or the other, so that this will be the one such premise, but the others will be affirmative.]

This is not another argument, and it does not seem to me to be presented as something which contributes to the matter at hand but rather, as it were, as a porism[292] that has come to light from what has been said. The theorem is quite worthy of discussion. It goes like this. He says that if a syllogism is negative, and it must establish premises by inserting terms for each, and then again [one must] establish the premises of that syllogism, and for the most part this occurs by thickening the premises with middle terms, then it is not as one might assume, that the affirmative as well as the negative premises increase in the same manner; rather only the affirmative [premises] increase, while the negative ones no longer do so. For in every negative syllogism, however far the premises are thickened, it is impossible for there to be more than one[293] negative premise. This seems to be paradoxical. For whenever a middle term is inserted in the negative premise the major [premise] is negative. For example, suppose that A belongs to no B. But B belongs to every C. It is pre-eminently clear that when the premise BC is thickened it produces only affirmative premises. But there is not yet a negative one [that does this]. For if I want to prove that A belongs to no B through the middle D, one must assume that A belongs to no D, but D belongs to every B; again, if you must prove that A [belongs] to no D through the middle term E, A must belong to no E, and so forth. How is it that he says that there is not more than one negation? I say that he was correct in adding 'in every syllogism'. For if we put together the premises as wholes and produce a single syllogism, only one negative premise will be found: A [belongs] to no E. E belongs to every D, D [belongs] to every C, C [belongs] to every B, therefore A belongs to no B.[294] So there is one negative [premise], and all of the rest are affirmative. So, every negative syllogism, whether simple or composite, has one negative premise, but the simple [syllogism] has one affirmative [premise] as well, while the composite [syllogism] has more than one affirmative [premise]. What occurs in the case of the first figure[295] occurs in the others, too. For when syllogisms are thickened, there will be one negative [premise], but the rest will all be affirmative.

86b27 If that by which we prove is better known and is more worthy of conviction, and the privative [demonstration] is proven through the affirmative [demonstration], [but the latter is not proven through the former, then it is better since it is prior and better known and more worthy of conviction.]

This is another argument. He says that that through which something is proven is better known and is more worthy of conviction than that item which is proven. And the negative is proven through the affirmative, but that one is not proven through this one, therefore the affirmative is better known and is more worthy of conviction and for this reason is also superior to the negative.

86b30 Another point is that if the universal immediate premise is the principle of deduction [and in the probative [demonstration] the universal premise is affirmative but in the privative [demonstration] [the universal premise] is negative, but the affirmative [premise] is prior to and better known than the negative [premise] (for the negation is known through the affirmation) …]

This is another argument. He says that immediate and universal premises are principles of deductions; the affirmative immediate [premise] is of the affirmative [deduction] and the negative immediate [premise] – let us suppose for the moment – is of the negative [deduction]. And I say this because not only is the immediate negation a principle of the negation but the immediate affirmation is as well. For now, let it be assumed that an immediate negation is a principle of the negative syllogism. For this [premise] is that which is significant for the conclusion, for the conclusion always follows from the inferior premise.[296] So if these are the principles of deductions and the affirmative [premise] is by nature prior to the negative [premise] and is better known [than it] (for if there is not an affirmative [premise] there will be no negative [premise] either). This is because the affirmative [premise] is a kind of possession, while the negative one is a privation, and a possession is prior to a privation, and if there is no possession there will be no privation of the possession. For a privation is a privation of a possession. But it is not also the case that a possession is a possession of a privation. So the possession is prior by nature and in time, and the privation of the possession is known through the possession), so that if the affirmative [premise] is prior to and better known than the negative [premise], and the demonstration that is on the basis of the [premises] that are prior and known to a higher degree is superior to that which is not on the basis of such [premises], it follows that the affirmative demonstration is superior to the negative [demonstration].

> **86b35** … and affirmation is prior, as being is to nonbeing, [so the principle of probative [demonstration] is better than [that of] privative [demonstration]. But the one using better principles is the better [demonstration].]

He says '[as] being is to nonbeing', that is [as] possession [is] to privation. For it is not at all possible to know anything about what is not. For we are not able to know the negation of this. For if we do not know what 'hippocentaur' or 'goatstag' mean, we will not be able to know the negation of this. So in such cases, I mean, of what does not subsist at all, affirmation must precede negation with respect to knowledge, in order for us to even know what it is that we deny.[297]

> **86b37** Further, it has the character of a principle to a higher degree. For there is no privative [demonstration] without [a demonstration that] proves.

This [means] the same as the lines prior to it. For he said above, too, that that by which something is proven is better known, and the affirmative [demonstration] is not proven by means of the negative one, but the negative one is proven through it. Therefore the affirmative [demonstration] has the character of a principle to a higher degree. And, again, there follows [the point] that the affirmative [demonstration] is prior to the negative one, as being is to nonbeing.

Chapter 26

> **87a1** Since affirmative [demonstration] is better than privative [demonstration], it is clear it is also better than [demonstration] that leads to the impossible. [But we need to know what the difference is between them. Well, let A belong to no B, and B to every C. Well, it is necessary that A belongs to no C.]

He has moved on to the third of the problems, that is, that direct demonstration is superior to [demonstration] to the impossible. And he now compares direct negative demonstration with [demonstration] through[298] the impossible.[299] For if direct negative demonstration[300] is shown to be superior to demonstration through the impossible, and affirmative is superior to negative [demonstration], then it is clear that *a fortiori* direct affirmative [demonstration] will also be superior to [demonstration] through the impossible.[301] 291,1

And since it is not possible to know which is superior, if you do not know how they differ from one another, let us first, he says, consider this very thing: what is their difference with respect to one another. Again, since someone who does not know what each of them might be at all, is incapable of knowing that [i.e. how they differ], for this reason he first expounds what direct demonstration is and what [demonstration] through the impossible. By the way, here, in the very exposition of these [kinds of demonstration], all commentators have attacked Aristotle en masse, saying that he expounds deduction through the impossible incorrectly. So we will first expound direct [deduction] and [deduction] through the impossible, as he has taught us [about them] in the second book of the *Prior Analytics*.[302] Then we will also expound what is here said by him,[303] and thirdly, [we will expound] the accusations, which the commentators bring forward with good reason against the things here said.[304] And afterwards, if we can, we will think of a defence on behalf of Aristotle.[305]

Now direct demonstration is [demonstration] through three terms, which, in the case of the first figure, predicates the major term of the middle term and the middle term of the minor term. For example, A belongs to no B, B to every C, therefore A to no C.[306] And in the case of the other [figures it works] in the manner described in the [chapters] about the three figures.[307] So that is direct demonstration. Whenever we want to demonstrate through the impossible that A belongs to none of the Cs, we assume the opposite of the conclusion to be true, I mean, of course, that A belongs to some of the Cs.[308] For if it is false that [A] belongs to none of the Cs, it is true that [it belongs] to some. So we assume that A belongs to some of the Cs, and we use that [assumption] as the minor premise. Then in addition we take another premise from outside, a universal affirmative one, for example, that D belongs to every A; but A belongs to some of the Cs: and it is inferred that D belongs to some of the Cs. Now let it be agreed upon that this is false and impossible, as we will also show with examples.[309]

So whence is the falsehood inferred, that D belongs to some of the Cs? Through the figure or through the premises,[310] and of the premises either through the major or through the minor or through both.
291,1 Now the figure is valid. But it is also supposed that the major premise, DA, is true – for we have assumed it to be true. Therefore, it remains that the minor [premise] is false. For as we saw it is impossible, if the figure is in good order, and both premises are true, for something false to have been inferred. So if by positing that A belongs to some of the Cs it is inferred that D belongs to some of the Cs, and it is false that D belongs to some of the Cs, it is therefore also false that A belongs to some of the Cs. And if that is false, it is therefore true that it belongs to none [of the Cs]. So the proposed [conclusion] has been proved through the reduction to the impossible, through two deductions, a categorical one and a hypothetical one, the

categorical one being a prodeduction,[311] and the hypothetical one inferring the very proposed [conclusion].[312] For the categorical one merely inferred that D belongs to some of the Cs. And the hypothetical one went forward in this manner: if it is true that A belongs to some of the Cs, it will also be true that D belongs to some of the Cs, just as the categorical deduction inferred. But the consequent is false, therefore so is the antecedent. Therefore, it is not the case that A belongs to some of the Cs. Therefore, it belongs to none, quod erat demonstrandum.[313]

Let the reasoning be made quite clear by means of examples. For let it be proposed that it be proven through the impossible that stone belongs to no human being. So I assume that stone belongs to some human being, then from outside I assume a universal and affirmative major premise, for example that lifeless belongs to every stone. But stone also belongs to some human being; therefore lifeless belongs to some human being. This is false and impossible[314] because of nothing other than the minor premise which says 'stone [belongs] to some human being'. So it has been inferred that, if stone belongs to some human being, then lifeless belongs to some human being as well. But the consequent is false, therefore so is the antecedent. Therefore stone belongs to no human being. Now if what needs to be demonstrated is negative, the premise included from outside needs to be universal, affirmative, and [the] major [premise]. For the opposite of the universal negative [proposition] is the particular affirmative, which will at any rate be given the place of the minor premise.[315] If the [thing to be] demonstrated should be affirmative, and if, on the one hand,[316] it should be particular, I mean of course that it belongs to something, the additional premise taken from outside is likewise affirmative, and it is certainly the minor premise and for that reason it does not matter whether it is universal or particular: the opposite of the particular affirmative, which is what we want to demonstrate through the impossible, is the universal negative. So that must be the major premise. Accordingly, the premise that is assumed in addition will be affirmative and the minor premise, as we said, and whether it is universal or particular is indifferent.[317]

If, on the other hand, the [thing to be] demonstrated is universal and affirmative, the [premise] assumed in addition from outside will again be universal and affirmative, but then it is not possible anymore to deduce it through the first figure. The opposite to the true [premise] [i.e. the universal and affirmative] is the particular negative, and this can be neither the major premise, because it is particular, nor the minor [premise], because it is negative.[318] So the categorical deduction is generated in either the second or the third figure.

Now it has been said what is direct demonstration and what is demonstration through the impossible. They differ from one another, because direct [demonstration] deduces the posterior from the prior (the conclusion is based on the premises), but instead demonstration through the impossible [deduces] the prior from the posterior; for by the conclusion being false, it also eliminates the premise. Direct [demonstration] inferred through A belonging to no B and B belonging to every C that A belongs to none of the Cs. [Demonstration] through the impossible, instead, produced a premise that is the opposite of that, I mean that A [belongs] to some of the Cs, and, as we have expounded, after inferring from this and another premise, DA (i.e. that D [belongs] to every A), that D belongs to some of the Cs, by the elimination of that conclusion it eliminated the premise as well. Accordingly it inferred the prior from the posterior. For the premises are prior by nature to the conclusion. And for this reason direct [demonstration] is superior to [demonstration] through the impossible, because the former proves the posterior from the prior, and the latter the prior from the posterior, and because direct [demonstration] deduces what it wants at the start, but [demonstration] through the impossible deduces the opposite of what it wants at the start; and after eliminating what it has concluded, it returns to what has been proposed [for demonstration] (and for this reason it is also called circular demonstration,[319] because it comes around full circle, as it were); and because the former is simple, but [demonstration] through the impossible is compounded of the categorical and the hypothetical. And even if sometimes direct [demonstration] also happens to require a hypothesis (for sometimes we are convinced of some of the premises through hypothetical deductions), that does not always happen.[320] But instead, the path of demonstrations through the impossible always goes through the categorical and the hypothetical.

So we have expounded deduction through the impossible and direct [deduction], on the basis of what is said by Aristotle in the second book of the *Prior Analytics*.[321] Here, however, Aristotle himself expounds them in a different manner. For after taking three terms, A, B, and C, and then expounding the direct deduction, [saying] that A [belongs] to no B, B to every C, therefore A to no C, he also expounds [deduction] through the impossible in reference to those same three terms. For let it be required, he says, to prove through the impossible that A belongs to none of the Bs, this time not by demonstrating the conclusion, which had to be proved directly, through the impossible as well, but instead [by proving] the major premise AB. So, he says, if it is not true that A belongs to none of the Bs, then it is true that A belongs to B. But B also belonged to C; therefore it is inferred that A belongs to C. Now let it be a matter of agreement, he says, that this very thing which has been inferred is false and impossible, not because B belongs to C (for that is taken to be true), but because A belongs to B. So if it has been inferred because of this that A belongs to C, and this is false, then it is also false that A belongs to B. Therefore it is true that it does not belong, quod erat demonstrandum.

So this is how Aristotle [expounds it]. The commentators accuse him, first, as I already said, of not having made the proof through the impossible with regard to the conclusion itself, but with regard to the premise. And second, they say, in what sense do you mean that A belongs to B: to some or to all?[322] For if [you mean] to some, then you render the deduction invalid, since the major premise is particular. But if instead you take 'belongs' to stand for belonging to all, then you do not infer what you want. For let A [belong] to every B, <B to every C>,[323] and A to every C. And suppose that this is false and impossible and is inferred due to the assumption that A belongs to every B. Now if by assuming that A belongs to every B it has been inferred that A belongs to every C, and this is false, then it follows that 'A belongs to every B' is false as well. Now if this is false, the opposite will be true, I mean that [A] does not [belong] to every [B]. But that is not what we wanted to prove: instead, [we wanted to prove] that A belongs to none of the Bs.

So these are the objections the interpreters with good reason bring against the argument. But we can defend it on Aristotle's behalf in this manner, [saying] that just as when he wanted to display to us the difference between the three figures in the first book of the *Analytics*,[324] he used different letters, A, B and C for the first figure, M, N, O for the second, and P, R, S for the third, so here, in order to display to us direct [proof] and proof through the impossible with the same terms, he did not take a different additional term from outside. For whenever we want to prove the same conclusion both directly and through the impossible, we do not take any additional term from outside in the [proof] through the impossible, but use the same [terms] that we used in the direct proof also in the [proof] through the impossible, changing only their order. For having taken the opposite of the conclusion and having added one of the premises to it, we thus eliminate the remaining [premise]. For example, let A [belong] to every B, and B to every C, and you will infer truthfully that A [belongs] to every C. This is how direct [proof goes], but [proof] through the impossible, instead, [comes about like this:] if someone does not agree that A [belongs] to every C, then it is clear that [A] does not [belong] to all [C]. So since A does not [belong] to every C, and B [does belong] to every C, it follows that A does not [belong] to every B. But it was assumed that [it does belong] to them all. So a falsehood followed, not because B was taken to belong to every C (for that is assumed to be true), but because A [was taken] not to belong to every C. Therefore this is false, and it is true that [A belongs] to every [C]. So this is the way in which the proof through the impossible uses the same terms as the direct [proof], when a deduction is generated through both. So this is why he did not change the terms.[325] When he assumes AB in the course of proving through the impossible, it is not because he employs it as a premise and wants to demonstrate it through the impossible but because he makes use of it as a thesis.[326]

So I mean that when he says that A belongs to B, he does not mean that it belongs to every [B], but to some.[327] For Aristotle says that simple propositions are equivalent to particular ones.[328] And he takes the [premise] that B belongs to every C not as minor premise but as major, in order that we think of it as transposed up. For he simply used the premise he has taken only for the sake of the example, not in order that the same order remain preserved for it, but, as I said, in order that we think of it as transposed. For it is not likely that Aristotle, the first and only one to have provided the logical methods, would commit such a grave error. Instead, as I said, he took the premises in this way only for the sake of the example; since, as we have said before,[329] he often also uses one letter instead of a premise.

296,1 **87a5** Now if we have assumed things in this manner, then the privative demonstration will be probative [of the fact that A does not belong to C].

'Probative', that is, the negative demonstration is generated directly.[330]

87a7 [This is how [demonstration] to the impossible goes:] If it were needed to prove that A does not belong to B.

See: he does not propose to prove the conclusion of the direct deduction through the impossible as well, but, as we said, [to prove] the major premise of the deduction (which is negative).

87a7 We have to assume that it does belong [and that B belongs to C, so that it follows that A belongs to C].

Note, again, that he did not say 'to every' or 'to some', but simply 'belong'. And we know that Aristotle says that simple premises are equivalent to particular [premises].[331] Next he links the premise BC, which is universal affirmative, to that [premise]. And he seems to take BC as minor [premise], but as I have already said, we need to understand it [i.e. the premise] not as staying the same in respect to its position, but it must be transposed upwards and placed higher.

87a9 Let it be known and agreed upon that this is impossible. [Therefore it is not possible that A belongs to B. So if it is agreed that B belongs to C, then it is impossible for A to belong to B.]

He does not take an indemonstrable hypothesis in general, but one that has credibility due to its self-evidence. For if someone sets out the syllogism with the terms, as we have done, he will find the impossibility of the conclusion. For it is possible to deduce something true from false premises, but not always:[332] and for this reason he also spoke as follows: 'let it be known and agreed upon that this is impossible'. For even if something true were inferred from false [premises], it would not be because of the premises, but because of the nature of the extreme terms. And if the conclusion were inferred because of the premises, then from falsehoods a falsehood would surely be inferred and from truths a truth.[333]

> **87a12** So the terms are arranged in the same way.

Having expounded direct deduction and [deduction] through the impossible, he then wants to say what they have in common and in what they differ. He says that 'the terms are arranged in the same way'.

> **87a12** The difference lies in which of the [privative][334] proposi- 297,1
> tions is better known, [that A does not belong to B, or that A [does not belong] to C.[335] Now when the conclusion is more known, i.e. that it is not the case, a demonstration to the impossible is generated ...]

He calls both the negative conclusion itself and the negative premise 'negative propositions'.[336] Now direct [deduction], he says, differs from [deduction] through the impossible, in respect to these negative propositions. For if the conclusion is more known than the premise, then a deduction through the impossible is generated since it infers the prior from the posterior. But if the premise is more known than the conclusion, a direct demonstration is generated, since it deduces the posterior from the prior.

> **87a16** ... but when the [premise] in the syllogism [is more known], then a demonstrative [demonstration] [is generated].

That is, when the premise which is taken 'in the deduction' is better known than the conclusion, a probative proof, that is, a direct one is generated, in the sense that it immediately proves what it proposes, while [deduction] through the impossible deduces that by the elimination of the opposite.

> **87a17** By nature the [proposition] 'A belongs to B' is prior to 'A belongs to C'. For the items on which the conclusion is based are

prior to the conclusion. [And 'A does not belong to C' is the conclusion and 'A belongs to B' is that on which the conclusion is based.]

Having said how they differ from one another, he then compares the proofs with one another, and shows that direct [proof] is superior to [proof] through the impossible, as the former deduces the posterior from the prior, while the latter does the exact opposite and deduces the prior from the posterior, just as the so-called proofs from signs establish the causes from the caused,[337] while natural demonstration goes in the other direction, and demonstrates the caused from the causes.[338]

87a20 For it is not the case that if something happens to be eliminated, then that is the conclusion, and those are the things on which it is based.

In order that no one says that [proof] through the impossible deduces the posterior from the prior, in saying 'if AB, also AC; but AC is false; therefore also AB is false' (for in the elimination he took AC as premise, but [he took] AB as conclusion), he therefore says '[it is] not [the case, that] if something eliminates something together with itself, then the latter is the premise of the former, and the former the conclusion of the latter; instead they are antecedent and consequent in hypothetical syllogisms, but they are not premise and conclusion'.[339] For of the premises, he says, one is major and the other is minor: but in hypothetical syllogisms antecedent and consequent do
298,1 not relate to one another in that way. For neither of them can be called either major or minor premise. A fortiori, then, the antecedent would also <not> be a premise of a syllogism.[340] And it [i.e. the antecedent] is not a syllogism, since a syllogism cannot be generated from one premise.[341]

87a22 But that on which a deduction is based is that which relates to it in such a way that it relates either as whole to part, or as part to whole.

'That on which it is based', that is, the premises. So the premises in a syllogism are those of which the major is a whole, and the minor a part thereof.[342] And it is clear that, when both premises are affirmative, the major is some whole, and the minor a part thereof. For the predicate is always of wider extent than the subject and encompasses it. But the same reasoning applies also when the major [premise] is negative: for in that case the middle term is coextensive with the major [term], since the universal negative [premise] converts. So if the middle term is of wider extent than the minor, and <in nega-

tions>[343] the major term is equal to the middle term according to conversion, then the major premise is of wider extent. Some <call the major premise more universal than the minor, even if the major term is more particular.>[344]

> **87a24** And the propositions AC and AB[345] are not related to one another in that manner.

That is, the antecedent and consequent in hypothetical syllogisms. He said 'propositions' instead of 'negations'.[346] So antecedent and consequent, he says, 'do not stand to one another in that manner' as the one being more universal and prior, and the other more particular and posterior.

> **87a25** If [a demonstration] from the more known and prior is superior, and both derive their credibility from something not being the case, but one from the prior and the other from the posterior, then privative demonstration will be better simpliciter than [demonstration] to the impossible[, so that it is also clear that the affirmative [demonstration], which is better than the [privative one], is also better than the [demonstration] to the impossible.]

This finally concludes the [matter] at hand. And he says that, even if both direct [demonstration] and [demonstration] through the impossible give credibility to what they want to give credibility to through a certain negation (for by assuming that A belongs to no B direct [demonstration] demonstrated that A does not belong to C either, and [demonstration] through the impossible did likewise by proving that 299,1
A does not belong to B), if, then,[347] each has credibility through some elimination, but direct demonstration [proves] the posterior from the prior, while [demonstration] through the impossible [proves] the prior from the posterior, then clearly direct [demonstration] will be superior to [demonstration] through the impossible. And if negative direct demonstration is superior to [demonstration] through the impossible, and affirmative [demonstration] is superior to negative, then obviously *a fortiori* direct affirmative demonstration will be superior to demonstration through the impossible.

Chapter 27

> **87a31** One science is more precise than another, and prior to it, if that same science studies both the fact and the reason why.

And with these words he again provides us with further theorems[348] useful to the theory of demonstration. First, which kind of 'science is more precise' than which. Well, he says that one science is said to be more precise than another in many ways.[349] In the first place, the science understanding the fact and the reason why is more precise than the one which knows only the fact. For instance,[350] the science that knows that the sun eclipses because of the moon standing before it, whenever this happens to it perpendicularly,[351] and that the moon, instead, eclipses because it falls within the shadow of the earth: this science is more precise than the one that knows that the moon has a spherical shape because it is illuminated in such a way.[352] For the first knows both the fact and the reason why;[353] but the latter, insofar as we know on the basis of the example given, knows only <the> fact (namely that it has a spherical shape),[354] obtaining its evidence from some attributes. For the phases are not the cause of its spherical shape, but of course only a by-product thereof. The cause of the moon having a spherical shape is either that it consists of the fifth substance, or perhaps that it is eternal, or some other such [cause]. So that kind of science is only of the fact.

The former knows the posterior from the prior, and *that* is knowledge of the reason why; but the latter knows the prior from the posterior.[355] That, then, is one way in which some science is more precise.

87a32 But not without the fact the reason why.

This should be read in an inverted order, as 'but not the fact, without the reason why'; for he who knows only the fact, without the reason why, does not know precisely.[356]

87a33 And the [science] that is not of an underlying subject matter [is more precise] than the one that is of an underlying subject matter, e.g. arithmetic [is more precise] than harmonics.

300,1 A second way in which one science is more precise than another. By 'of an underlying subject matter' he does not mean the universal. In that case the result would be the opposite, for he does not want the science of the partial and the particular to be more precise than science about universals.[357] By 'not of an underlying subject matter' he here means the intelligible and immaterial, as he made clear through the examples given, and by 'of an underlying subject matter' [he means] the sensible and material. So the science, he says, which concerns the immaterial and the intelligible is superior to that which concerns material and sensible things. For this reason arithmetic is superior to and more precise than harmonics: the one studies the numerical ratios themselves in themselves, whereas the other [studies the numerical ratios] in the strings [of an instrument]. Likewise, geometry is more precise than optics, because the former studies the accidents of the figures and lines themselves in themselves, without any matter, but optics [studies] the accidents of the rays of vision that are shaped in some way or another.

> **87a34** And the one [that starts] from fewer [things is more precise than the one] [that starts] from an addition, e.g. arithmetic is more precise than geometry.

This is a third way in which one science is more precise than another. For the [science] that is about things that are more simple, he says, is more precise than the one that is about things with a higher degree of composition. For that is what 'from an addition' means.

For example, the spherics of Theodosius is a more precise science than that of Autolycus, which treats of the moving sphere. For the one simply looks into the accidents of the sphere, without considering in addition whether it moves or not. But Autolycus studies the accidents of the moving sphere. In the sciences additions always make the subject more particular and for this reason less precise. And yet, Autolycus' science of the moving sphere is more precise than astronomy; for that, finally, studies the moving sphere *with* matter. For it studies *this* moving [sphere], I mean the heavenly [sphere]. And for this reason it lacks precision. So everything that is proved in astronomy does not offer the utmost precision, but the approximate. For example, they say that the sun stands from the moon more than eighteen times, but less than twenty times, the distance that the moon stands from the earth; for in these things we should be satisfied with approximating precision. And for all other things that are proved in astronomy the same argument holds. So that the spherics of Theodosius should not be taken as additional elements of astronomy, but as principles and causes of the things that are demonstrated in astronomy. For through them, as causes, those things are demonstrated. And the relative position that arithmetic has to harmonics and geometry to optics, that same relative position Theodosius' spherics has to the study of the sphere in motion, and the latter to astronomy. For the higher sciences are always the causes of the lower ones.

> **87a35** I mean by 'from an addition', for example, that the unit is a substance without position, whereas the point is a substance with position; the latter [results] from an addition.

Since he mentioned arithmetic and geometry as examples of [sciences] '[starting] from fewer things' and of [sciences] '[starting] from an addition', arithmetic as more simple, and geometry as more composite, through these words he has added in what sense this is true. Saying 'I mean by "[starting] from an addition"', he first states in what sense arithmetic is more simple, and then also how geometry [starts] from an addition. Now, he says that arithmetic uses more simple principles. For arithmetic uses the unit as principle. Likewise, the geometer uses the point, which is itself also a kind of unit. But the arithmetician takes the most simple [principle], the unit, while the geometer takes the unit which is situated somewhere. And it is more simple to take the unit without qualification than it is to take the unit that is situated somewhere. And saying 'unit without position' does not involve 'without position' as an addition, as someone might suppose. For it is not the case that, just as in saying 'unit with position', 'with position' is an addition, so too in the case of saying 'unit without position', 'without position' is an addition. For 'without position' is nothing other than a privation and negation, and obviously not a positing.[358] And, when he said 'I mean by "from an addition"', and first gave the example without addition, saying 'for example, that the unit is a substance without position', and subsequently in the second place [gave the example] of 'from an addition', saying 'the point is a substance with position', he added 'the latter [results] from an addition' – i.e. I mean that the point [results] from an addition, to prevent people from getting confused.

He calls the unit a substance, following the Pythagoreans, about whom we have often said that they symbolically indicated the kinds and the natures and essences of things with numbers.[359]

Chapter 28

302,1 **87a38** A science is one if it is of one genus, i.e. all things that are composed of the first things and are parts or per se attributes of them.

He has moved on to another theorem, namely when something is 'one science'. For some people might think that even if there is not one science in species, there is still one in genus, such as arithmetic and harmonics. On the other hand, geometry and stereometry might still seem to be different [i.e. in species], even if there is no difference [i.e. in genus]. Likewise with optics and geometry. For this reason, he teaches us guidelines, by which we should judge what is one science and what is not one. So he says that a science is one if it concerns one genus and investigates what holds of it per se. And sciences concern one genus if they use the same principles. For if the genus is one, the principles would be the same as well, and if the principles are the same, then the genus would necessarily be the same as well. For example, geometry and stereometry and optics are one science: they have common principles and elements of which their demonstrations are composed; and the genus of them all is one, I mean of course the continuous, or magnitude, and the species thereof. So those sciences are the same. He says that the scientific theorems[360] 'are composed' of the proper principles, because the theorems are generated by the combination of the principles. For example, lines, points, planes, angles, circles and the like are the elements of the geometer; and these, by being successively combined with one another, make the scientific theorems in geometry.[361] For the triangles and parallelograms are composed of straight lines and angles, and the rest in the same way. He says 'parts and attributes of them', i.e. of the principles. Parts are for example the segments of the semicircle, the triangles made by cutting four-sided figures across the diagonal, and attributes are things such as bending, being deflected,[362] meeting and the like. The general thought and the construction of the text are something like this: a science is one if it concerns one genus of theorems;[363] theorems are about one genus if they are composed of the same first principles. Things belong to one genus also if they are 'parts and attributes of them'. The words 'of them' either mean 'of the theorems' or 'of the first principles'.[364] Now if he means 'of the theorems', he would say, for example, of the fifth theorem that a part of it is 'that the angles at the base of the isosceles triangles are equal to one another', while the whole theorem also shows that the [angles] below the base are [equal].[365] An attribute is, for example, that the [angles] below the base are greater than the ones at the base.[366] If, on the other hand, he means 'of the principles', it was in order that we might conceive of parts [such] as triangles delimited by the diagonals of foursided figures or the segments of the circle, and attributes, as I was saying,[367] the assumption of straight lines that bend or angles that are such and such, and things like that.[368]

> **87a39** One science is different from another, if their principles neither consist of the same things, nor the [principles][369] of the one science consist of those from the other.

For if the sciences that use the same principles are the same science, it is clear that different sciences would be ones that do not use the same principles. He added very precisely 'nor the one from the other', for if they do not use the same principles, but one of them uses the theorems of the other as principles, then they are also the same science, such as is the case with geometry and optics. For optics, taking as principles the things that have been demonstrated in geometry, thus proves its own theorems. So these [sciences are] the same [science].[370] So for sciences to be different [sciences] it is necessary both that they don't use the same principles and that one does not use the theorems of the other as principles.

> **87b1** Evidence of this is found when one reaches the indemonstrable things.

It is clear, he says, that those sciences we mentioned are indeed different, when you ascend to their respective indemonstrable principles. For in the sciences that are different the [indemonstrable principles] must not have any kinship with one another. In the case of optics, after all, if we wanted to ascend to the indemonstrable principles, we would come to the same things that we would come upon when we ascend from the theorems of geometry. So all those [sciences] of which the indemonstrable principles are the same belong to one science, and all those of which the indemonstrable principles have nothing in common, are different sciences.

> **87b2** For the [undemonstrated principles] have to be in the same genus as the things that have been demonstrated.

For the genus the triangles and parallelograms and circles and the like are in (they are in a species of quantity, namely the continuous), that is the [genus] the undemonstrated elements, on the basis of which geometrical [theorems] are demonstrated, are in as well. If both the principles and what follows from them are in the same genus, then it is clear that those sciences of which the first principles are not the same, are different, if, indeed, that science is one which concerns some one genus. For this reason, geometry is different from arithmetic, because the principles are different. Of the one they are points, lines, planes and the like, and of the other they are units and uneven and even numbers and the like. And the former are in the species of the continuous, the latter in that of the discrete.

> **87b3** Evidence of this is [found] when what is demonstrated through them is in the same genus and akin.

He said that evidence of sciences being different is the fact that the indemonstrable principles are different. Again, he says that evidence of the indemonstrable principles being different is [the fact] that what is demonstrated through them is not in the same genus.[371] For example, the [facts] of arithmetic and those of geometry are in the genus of the discrete and of the continuous respectively. For this reason their principles are different, too. And this is to be expected: for the principle is a principle of certain things,[372] and these are among the [items in the category of] relation.[373] So if what follows from a principle is different, the principles must be different as well, and if the principles are different, what follows from them must be different as well.

Chapter 29

> **87b5** It is possible for there to be multiple demonstrations of the same thing, not only by taking a non-continuous middle term from the same series [of predication] …[374]

He has moved on to another topic, namely whether it is possible to demonstrate the same thing through multiple demonstrations or whether there is [just] one demonstration of each demonstrable [fact]. And it is clear that the same thing may be demonstrated in multiple ways, because it is possible to demonstrate the same thing not only both directly and through the impossible, but also several times with each of these methods. This is clear from the minor assumptions:[375] they are put forward with more than one [method], and each of them demonstrates the same thing, now from this angle, now from that. For it is possible to find more than one middle term affording demonstration of the same thing, and not just more than one middle term, he says, 'from the same series', but also now from this [series], now from that. With 'from the same series [of predication]' he means the subordinate middle terms. For example, if we wanted to prove that human being is a substance, it would be possible to prove it, both through the middle term 'body', and through the middle term 'animal' and through the middle term 'rational'; now these are from the same series. It is also possible to prove it from a different series, if we take as middle term 'two-footed' or 'walking upright' or 'conversing' or something like that. For these are not from the same series as the ones [mentioned] earlier. So then, he asks, if the middle terms, from which the same thing is demonstrated, [can also be] from different series [of predication], do they have nothing in common with one another, but is it necessary that these are denied of those,[376] or is it necessary that they do have something in common? And he says that it is necessary that they do have some kind of association: for combined with the subject term in the thesis or the conclusion they produce the third figure [syllogism],[377] so that it is necessary that they are also partially associated with one another. For if, for example, man is an animal, but man is also two-footed, then you will infer that there is some two-footed animal.[378] And the same account [holds] for all.

> **87b7** Such as C and D and F of AB.[379] [But also from different [series [of predication]]. For example, let A be 'to change', and let D stand for 'to be in motion', B 'to feel pleasure', and again G 'to come to rest'. So it is true to say both D of B and A of D. For he who feels pleasure is in motion and he who is in motion changes. Again, it is true to say A of G and G of B. For everyone who feels pleasure comes to rest and he who comes to rest

changes. So the syllogism [is generated] through different middle terms which are not from the same series [of predication].]

He takes AB as the thesis, B as the subject term, A as the predicate, C, D and F as the middle terms from the same series [of predication], and G as the middle term from a different series. He calls 'to change' A and 'to feel pleasure' B. Now it is possible to prove that what feels pleasure changes, both through the middle term 'motion' and through the subspecies of motion. For example, 'what feels pleasure moves, what is in motion changes', therefore what feels pleasure changes. Again, what feels pleasure is altered, what is altered changes, so what feels pleasure changes. And being altered is adjacent to feeling pleasure, and being in motion to being altered.[380] It is not only possible to demonstrate the same thing through [middle terms that are in] the same series but not adjacent, but also through [those that are not in] the same series, such as 'to be brought to rest':[381] for 'that which feels pleasure is brought to rest, that which is brought to rest changes', therefore that which feels pleasure changes. In the fifth book of the *Physics* he says that being brought to rest ends up at rest itself.[382] For being at rest is one thing, such as the very state of being motionless, and being brought to rest is another. For that is the end of motion.[383] So being brought to rest is intermediate between motion and rest. It is clear that pleasure occurs especially in that part of motion [i.e. in being brought to rest], in all of those motions to which feeling pleasure belongs.[384]

87b14 But not, of course, in such a way that neither of the middle terms is said of the other. [For they must both belong to some one thing.]

Since the middle terms of one and the same thesis may be now from one, now from another series [of predication], they must also partially hold of one another on account of how they, together with the subject [term], produce the third figure [syllogism], as we said. So surely, this is also how it is in the case of the [text] before us: for being brought to rest belongs to something that moves, and something that moves is brought to rest.[385]

306,1 **87b16** To be investigated: in how many ways it is possible to generate a deduction of the same thing through the other figures.

Since he developed the argument with respect to the first figure, it is fitting that we investigate also in 'the other figures in how many ways it is possible to generate a deduction of the same thing'. It is clear that also in the other figures it is possible to demonstrate the same thing both through middle terms that are in the same series [of predication] and through middle terms that are not in the same series.[386]

Chapter 30

87b19 There is no scientific understanding through demonstration of [what comes about] by chance. [For that which is by chance is not necessary or for the most part, but [it is] what happens besides these things. But demonstration is of either one of the former two. For every deduction [comes about] either through necessary premises or through [premises that hold] for the most part. And if the premises are necessary, then the conclusion is necessary as well, and if they are for the most part, then the conclusion is such as well. Accordingly, if what [comes about] by chance is neither for the most part nor necessary, then there will be no demonstration of it.]

He shows in a summary fashion that 'of things that [come about] by chance there is no' demonstration – what am I saying? demonstration? no deduction at all: for every deduction, he says, either has necessary premises or premises that belong to 'what holds for the most part'.[387] 'And if the premises were necessary, the conclusion would be, too'; if the premises were 'for the most part', then the conclusion would be like that too. But that which comes about by chance is neither necessary nor for the most part, but belongs to the things that are in very few cases.[388] So it is impossible to know that which comes about by chance through a deduction. For no deduction has premises that are true of very few cases, but, as I said, either always or in most cases. So we only know that which comes about by chance through perception, but not through a deduction.

Chapter 31

87b28 It is not possible to have scientific understanding through perception.

The matter at hand is to prove that perception is not scientific understanding. For someone might think perception and scientific understanding to be the same thing in this manner:[389] if, after all, both perception is capable of apprehending qualities (colours, and sounds, and the like) and scientific understanding concerns quality (for geometry is about figures and their per se accidents,[390] and music is about the sounding of sounds; and moreover magnitude and motion and number belong to the common sensibles,[391] with which astronomy and arithmetic are also concerned), well, if perception and scientific understanding are about the same things, then how could they not be the same? So this is the puzzle he solves when he says that, even if perception and scientific understanding are about the same things, [they are] not [about them] in the same manner. For perception apprehends particular things: it does not know every 307,1 white thing without qualification or every circle, but every case of 'this particular thing', the particular that is in some place and subsists now.[392] Scientific understanding, however, knows about every white thing without qualification, regardless of where it might be, and not that which is now, but all that which has become and [all that] which will ever be.[393] Accordingly, if one knows the particular that is situated somewhere and exists now, but the other the universal and that which always is, perception will not be scientific understanding, nor will the object of perception be an object of scientific understanding. For nothing particular is an object of scientific understanding. And besides, even if they both concern qualities, they are not about the same [thing], but perception is about that which holds accidentally, and scientific understanding about essential quality, which cannot be apprehended by perception. For that the three angles of a triangle are equal to two right [angles], or the ratio that the strings have to one another when they are in harmony and similar things, does not belong to perception to know, but to reason, and distinguishing essential quality from accidental [quality] does not belong to perception, but to scientific understanding. That quality is not only accidental but also essential, has also been said in the *Isagoge*, because 'the differentiae and the species determine the quality relating to essence'.[394] We could say the same thing also about quantity.

Then, having shown only for the sake of completion that perception is not scientific understanding,[395] he says that, even if we were to perceive those things of which there is in fact scientific understanding, we would not be having scientific understanding of them. For example, if we had seen, he says, that the three angles of a triangle are equal to two right [angles], we would have neither a demonstration nor scientific understanding of it. For we would not know it without qualification, if we had seen that this [is the case] of this triangle here. For it has been said that demonstration is of universals and of what always holds. For example, even if 'we were on the moon', he says, and we were to see it being blocked by the earth and for that reason undergo an eclipse, we would not yet have scientific understanding and a demonstration of the lunar eclipse, but we would only know that this particular eclipse happened because of the blocking by the earth; of course we would not yet also [know] that every eclipse occurs in this way. For perception does not apprehend the universal. And scientific understanding [comes about] by 'knowing the universal'.[396] For even if we induce the universal from this, namely when by perception we often see the same thing occur, nonetheless the perception is not in addition [a kind of] scientific understanding. For it is not [perception] itself that has apprehended the universal, no: it attended to one and the same particular many times, but of course reason deduced the universal from them, while perception was fulfilling the role of a tool for reason. So it is clear that perception is not scientific understanding. For the former, as I said, knows only this thing, which is present and seen, and [it] only [knows] that it is, but not also the explanation. For even if we have seen that the moon eclipses on account of the blocking by the earth, and [we have seen] this very thing, I mean that the blocking is the cause, nonetheless it is not perception but reason which has deduced this, while perception only knows the fact. So for this reason he says that the universal is also more valuable than perception, because the universal reveals the explanation, while perception is unaware of the explanation. He says that such a universal, I mean the one with the explanation, is not only more valuable than perception, but also [more valuable] than unqualified 'intellection'[397] without the explanation. For example, if someone were to grasp with unqualified intellection that the triangle has internal angles equal to two right [angles], he would not know the explanation; intellection *with* the explanation is more valuable than this one. So then is intellection with the explanation superior to any unqualified intellection without explanation? Not at all. So for this reason he correctly added 'all things the cause of which is different',[398] i.e. whenever in one and the same intellection the cause is something other than the fact. For in these cases the [intellection] with the explanation is superior. Because if it is not possible to know the other sort of things with the explanation because there is nothing at all which has the character of a principle more and is a more important cause than they are, then the definition of intellection of these things is different.[399] After all, it is not the case that, since we know the common notions[400] without an explanation, the knowledge we have of them is inferior to the kinds of knowledge which are accompanied by an explanation. For such knowledge [i.e. of the common notions] is superior to scientific understanding. So we speak in the same way also in the case of knowledge of the first cause: for the knowledge concerning that is superior to any scientific understanding.

In addition to these things, he raises the question concerning what we meant earlier, if indeed perception is not scientific understanding, when we said that if some perception is lacking, some scientific understanding must also be lacking.[401] For these things are not mutually consistent: for if, when some perception is lacking, some scientific understanding must also be lacking, it seems that perception is scientific understanding. So he says that, even if, when some perception is lacking, some scientific understanding must be lacking as well, still it is not the case that, because of this, perception is scientific understanding. For there would not be scientific understanding without perception, just as there would not be a harmony without a lyre and without the concord of high and low sounds. But of course it is not the sounds that are the harmony but the ratio belonging to the art, by which the harmony is conceived. Likewise, then, no scientific understanding occurs without perception, but of course it is not the knowledge belonging to perception that is scientific understanding, but intellect which deduces the universal from the perceptions.[402] For it belongs to perception to perceive this white and again another [white] or even many times the same [white], and in this way, from the things that are known many times by perception, intellect deduces universal white without qualification. And likewise in every case. So if intellect deduces the universal from the particulars, then, as is to be expected, since the principle which knows the particulars is lacking, the deduction of the universal is lacking as well. And that [i.e. the deduction of the universal] is scientific understanding.[403] He subsequently uses an example such as this. We now investigate, he says, how light goes through lanterns, I mean the ones made of glass[404] or horn, and things like that. And some say that, because there are fine holes in the glass or the horn that are invisible to us on account of how fine they are, the light goes through them and in this way the air around them is illuminated through them.[405] Others do not think this to be true (for in that case it would be necessary, they say, not that the air is continuously[406] illuminated by them [i.e. the pores], but that the areas [of the lanterns] that do not have holes are dark), rather [they think] that the transparent body is the cause of its activity, i.e. light, being spread.[407] So he says that, if we had seen that the glass had pores and that the rays went through them, we would no longer wonder how the light inside went through, but we would clearly understand that light universally speaking goes through all glass like that; but we would of course not say that seeing that this light went through this glass here was scientific understanding, but that, from seeing that many times, reason inferred that this happens also in the case of all glass, even if it is not seen (after all, perception knows only what is present and seen); but now, since we do not see this occurring in this manner, we don't have scientific understanding of it at all – as in that case, likewise we also say in general that, in all cases of lacking perception, scientific understanding is lacking, but not, of course, that perception and scientific understanding are the same.

87b28 For even if perception is of the so and so and not of this particular thing, it is still necessary to perceive this particular thing and somewhere and now. [But it is impossible to perceive the universal and what holds in every case, since it is neither a

> particular thing, nor now. For then it would not be a universal, since we say that that which is always and everywhere is universal. So since demonstrations are universal, and we cannot perceive them, it is clear that it is also not possible to have scientific understanding through perception.]

That is, even if perception apprehends the quality (for there is no perception that it is a stone, but that it is white or that it is hard or that it is heavy or something like that, which scientific understanding apprehends as well), then at all events, he says, one still perceives this particular thing, that is, the particular and what is in that place (for that is what 'somewhere' means) and what is now, but not everything which is proper to perception, without qualification. Of course, scientific understanding does not apprehend the particular nor what has a position somewhere, nor what is now, but what is 'everywhere' and 'always'. For that is what the universal is like. Perception and scientific understanding, therefore, do not apprehend the same things. So that if perception does not apprehend universals, but scientific understanding [is] of universals, perception will not be scientific understanding and the object of perception [will not be] the object of scientific understanding.

> **87b35** But it is clear that even if it were possible to perceive of the triangle that it has angles equal to two rights etc. [we would seek a demonstration and we would not have scientific understanding, as some say. For perceiving is necessarily [of the] individual, but scientific understanding is to become acquainted with the universal.[408]] 310,1

Some say that, if we had seen 'that the triangle has three angles equal to two rights', we would have had scientific understanding through perception. So he says that, even if we had perceived such a thing, that would not be scientific understanding: for we perceive this triangle, but of course not every [triangle]. And scientific understanding is knowledge of the universal.

> **87b39** And for this reason even if we were on the moon and saw the earth blocking the light of the sun, we would not know the explanation of the eclipse. [For we would perceive that it is now eclipsed, and not why [it eclipses] at all. For as we saw there is no perception of the universal. Nevertheless, we would possess a demonstration if we sought the universal from seeing it happen many times. For the universal [becomes] clear from several particulars.]

That would be the case, clearly, if we perceived the universal eclipse.

That this is what he says, is clear from [the arguments] he brings forward: 'for we would perceive', he says, 'why it eclipses now, not why [it eclipses] at all; perception is not of the universal'. Of course, he says, we inferred the universal through the reason of perception by often seeing the same thing occur,[409] and thus we would possess scientific understanding again, just as now, having seen that this white here disperses the rays of vision,[410] and another and another, we have inferred that therefore also all white is like that.

> **88a5** The universal is valuable, because it reveals the cause. [So concerning those things, of which the cause is something other [than the fact], the universal is more valuable than perceptions and than intellection. But concerning first principles there is a different account.]

For when I say 'all fire is warm', I did not say the reason why, but only the fact. When, however, I say 'every eclipse of the moon occurs on account of a blocking by the earth', I have said the cause. Whenever [I say] as follows 'every eclipse of the moon is a privation of the light of the sun', I have said only the fact, but not, of course, the reason why and the explanation. So the universal is in general more valuable than perception, since the whole is more valuable than the part, and all intellection is universal in general:[411] for intellection is not of the particular and of individuals, even if some cases of intellection are more universal than others. In these cases, then, one is [the fact itself], and the other its cause, and the intellection with an explanation is superior to the one without explanation. In the case of those facts of which there is no cause, but they are the first and the most important causes of the others, intellection of these things is most valuable: for it is the intellection of the explanation of everything.

> **88a10** [So it is clear that we cannot have scientific understanding of anything demonstrable by perceiving it.] Unless someone would call that perceiving, i.e. having scientific understanding through demonstration.

That is, unless the one who is speaking employs the term 'to perceive' in regard to understanding: since perception in the strict sense would not be scientific understanding.

311,1

> **88a11** However, in the case of problems there are some instances that are attributed to the lack of perception. [For some we would not seek, had we seen them. Not, obviously, because we would know them by seeing them, but because we would

> possess the universal from seeing them. For example, if we had seen that glass[412] has holes in it and that light goes through [them]…]

He added this in order to solve an apparent contradiction. For in the foregoing, he said that, if perception is lacking, scientific understanding is lacking as well. In that way, perception would seem to be scientific understanding, which he now <refutes>.[413] So for this reason, in order to solve that [problem], he takes up the argument again. For it seems, he says, that some problems, concerning which scientific understanding has no apprehension, can be attributed to 'the lack of perception', that is, the reason for our ignorance concerning them is the fact that the apprehension of them by perception was lacking. For example, people puzzle about how light goes through glass, and some say that light goes through the pores of the glass, while others say otherwise. Now if the pores had been seen, science would not have puzzled about this. But even if the lack of perception is the explanation of the lack of scientific understanding, still perception is not the same as scientific understanding, as has often been said.

> **88a15** It would also be obvious why it burns.[414]

That is, 'it is visible'. If the pores had been seen, he says, the reason for the sun or fire shining through the glass would have been obvious.

> **88a16** Because[415] although for each [piece of glass] [we] see separately, still we understand that it is like that for all cases taken together.[416]

That is, perception, as we saw, perceives only the particular, but intellect, as it collects into a unity the [instances] that have been seen separately by perception, produces the universal and scientific understanding.

Chapter 32

> **88a18** That it is impossible for the same things to be principles of all deductions [they first study formally. For some deductions are true, and others false. And even if it is possible to deduce a truth from falsehoods, this occurs only once.]

The matter at hand is to show that the principles are not the same for all sciences. It is clear that the 'principles' [here] are the premises of deductions. Some of the premises are proximate,[417] others are first 312,1
[premises], from which the proximate ones also derive. So he proves that neither the proximate, nor the first principles of deductions are the same for all deductions. For now, he shows that the proximate

ones are not the same [for all]. And 'first they study it formally': he says 'formally' instead of 'more generally, applied to all deductions whatsoever'. Then he will also prove separately with regard to scientific deductions that it is not possible for the premises of all sciences to be the same. So what does he say here? That if 'some of the deductions are true and some false', and [if] it is impossible for a false conclusion to be inferred from true premises, but of a false conclusion the premises must surely also be false,[418] while of a true conclusion the premises are surely true,[419] and [if] the false is by nature distinct from the true and they are different from one another, then therefore it is not possible for there to be the same principles for all deductions. So what does he say? Is it not possible to draw a true conclusion from false premises, too? For I say 'human being is a stone, stone is an animal, therefore human being is an animal'. Look! The premises are false, but the conclusion is true. But surely from false [premises] a false conclusion is drawn as well. So it is possible [to draw] both a true and a false conclusion from the same [principles], inasmuch as they are true and false. Now he says that, even if it is possible to draw a true conclusion from false [premises], 'this occurs only once'.[420] Since surely, if we wanted to deduce each of the false premises, it would be altogether necessary to deduce them from false premises: for it is impossible that true premises lead to a false conclusion. Now of course, true conclusions from true premises do not admit of this just once, i.e. of true things being inferred through true things, but however much you would thicken the deduction,[421] always demonstrating the premises, the same thing happens, I mean that you also obtain true premises of the true conclusions.[422]

> **88a22** For example if it is true that A holds of C, while it is false that the middle term B holds. [For neither does A belong to B, nor does B belong to C. But if middle terms are taken for those premises, they will be false, because of the fact that every false conclusion depends on false [premises], and true conclusions on true [premises], and falsehoods and truths are different.]

A is the predicate, C the subject, and B the middle term which is falsely combined with both so that two false premises are generated. Now when, he says, middle terms are assumed for these premises, for example the middle term of AB and of BC, which are false, the premises which lead to them must also be false: for a false [conclusion] is drawn from false [premises].

313,1 > **88a27** Next, not even falsehoods [are inferred] from [principles] that are the same as each other. [For there are falsehoods which both are opposite to one another and cannot be [the case] at the same time.]

Not only, he says, are the principles of deductions different for this reason, because some are [inferred] from true premises and some from false [premises], but also those very deductions which consist of falsehoods have different principles.[423] For falsehoods are not the same as one another. For some falsehoods are contrary,[424] some are things that cannot possibly both hold, some are neither contrary nor simply impossible, but, while they can both hold, they nonetheless do not both hold.[425] For example, when I say that justice is injustice, or that the just is unjust, or that moderation is intemperance, I have said a contrary falsehood.[426] But when I say, 'human being is a horse' or 'stone is a stick', I have said an impossible falsehood. For these things are not contrary to one another but it is merely impossible that they both hold of each other.[427] And when I say of Socrates, when he is sitting down, that he is not sitting down, or when he is walking that he is not walking, I have neither said a contrary falsehood nor an impossible one, but a possible falsehood. Well now, if 'falsehood' [can be said] in more than one way, and it is impossible that these [different kinds of falsehood] are the same as each other,[428] then it is obvious that it is impossible that the principles of false deductions are [all] the same. So if even the principles of false deductions are not the same, then obviously they are far from the same for all deductions.

88a28 For example, that justice is injustice or cowardice[, and that human being is a horse or a cow, or that the equal is more or less.]

One of these is a contrary falsehood, namely that justice is injustice, and the other an impossible falsehood, namely that justice is cowardice, just as, in fact, 'human being is a horse or a cow' is impossible. Likewise also 'the equal is more' or 'is less' is impossible and contrary, at least as long as I take 'equal' without qualification.[429]

88a30 From the present [deductions] we argue as follows. [The principles are also not the same for all truths. For the principles of many of them are different in kind, and do not even apply [to one another], for example, units do not apply to points. For the former do not have position, while the latter do.][430]

He has shown 'formally', as he himself said,[431] that the principles are not the same for all deductions ('formally' because this holds for all deductions whatsoever, both true ones and false ones). Now he wants to work through the argument only with regard to true deductions, or rather, demonstrative ones.[432] For, he says, 'from the present deductions',[433] I mean the demonstrative ones, we will prove that, i.e. that it is not possible that the principles are the same for all such deductions. He shows it with several dialectical proofs, namely that (1) neither can the proximate principles of demonstrative deductions be the same, (2) nor can the first [principles].[434] And for now, he shows in the following manner that the proximate principles are not 314,1 the same for all deductions: for if the premises from which the conclusions or even the problems are derived are akin to the conclusions, and the problems are not the same as one another but many of them also differ in genus (for geometrical theorems differ in genus from those of arithmetic; for the latter concern discrete quantity, the former continuous [quantity].[435] Likewise they differ from musical [theorems]; for the latter are about the sounding of low and high sounds and their commensurability, and these things belong to quality, but those to quantity.[436] Likewise there is nothing that all the things mentioned have in common with the problems of medicine; for the problems of medicine study the substances themselves, with their matter, but the others concern either quantity or quality)[437] – now if the problems differ, and the premises are akin to [and of the same genus as][438] the problems, surely the premises must also be different in genus. So they are different in this sense, he says, that they do not even apply to one another, i.e. it is not possible to predicate affirmatively one of the other. For that is what 'applying' is. For in arithmetic the principle is the unit, and in geometry the point; and these things do not apply to one another, since the latter has position and the former does not have [position].[439] Likewise the principles of arithmetic are [for example] that of the numbers some are odd and some are even, and in geometry that the line is a magnitude in one dimension,[440] and the surface [a magnitude] in two [dimensions]. How would these things apply to one another or be predicated of one another? I could say the same thing with regard to medicine, and music, and astronomy, and [all] the other sciences. Therefore even of the true deductions, then, the principles are not the same.

88a34 But they must apply[441] either in the middle[442] or from above or from below, or else some of the terms must be inside and some outside.

After he has said that they do not apply, he proves this, saying what he takes 'applying' to mean here, other than 'the one being predicated of the other' or 'one being the subject [of the other]'. So [a term][443] either 'applies in the middle', he says, as in the case of the first figure, so that it serves as the subject of one of the extremes, namely the major, and is predicated of the other, namely the minor; or 'from above', as in the case of the second figure,[444] so that it is predicated of both extremes; or 'from below', as in the case of the third figure, so

that it serves as the subject for both extremes.[445] In the *Prior Analytics* he specified the middle terms in the same way.[446] [People][447] 315,1
also investigate in that [context] in what sense he says that the middle term applies to the extremes in the second figure. For if we take two affirmative premises, so that the middle term also applies to the extremes, the conjunction becomes invalid.[448] For a valid deduction to be generated, one of the [premises] needs to be negative. So how can he say that they apply to the extremes 'from above'?[449] We say that he is not here concerned with teaching about valid or invalid conjunctions, nor with expounding a valid conjunction in general, but only teaches us in how many ways a term can apply to [other] terms – surely as yet there is no combination of premises whatsoever.[450] And that comes down to saying that he means by 'applying' 'the one being predicated in general of [the other] or being the subject for the other, either affirmatively or negatively'. Perhaps we should even understand 'they must apply <either> in the middle or from above or from below' in a more simple manner, that is, either middle or major or minor terms must be used;[451] for that is what 'from above' refers to ([they do] not [apply] from above [to items] in a syllogism, but [to items] in a conclusion) and again 'from below' [is understood in the same way]. The phrase 'or else some of the terms must be inside and some outside' is either, pleonastically, the same as what [came] before (for in the <*Prior*>[452] *Analytics* he also said that the middle terms of the second and third figure have a position on the outside, but the ones of the first figure [have a position] on the inside, because they have a position adjacent to the extremes),[453] or by 'inside' he means the subject terms, and by 'outside' the predicates, that is, [he means] that the terms that are said to 'apply' either serve as the subject to those they are said to apply to, or are predicated of them. He calls the subjects inside and he calls the predicates outside, because the predicates encompass the subjects, and what encompasses, being outside, encompasses what is encompassed, keeping it inside itself.[454]

> **88a36** But it is not even possible that among the common principles there are some from which everything can be proved [(by the common [principles] I mean for example that everything is affirmed or denied), for the kinds of beings are different, and some things belong only to quantities, others only to qualities, and things are proved through the common [principles] with these things.[455]]

Having proved that the proximate principles are not the same for all sciences, he now proves that the universal and common notions are also not entirely the same for every science.[456] For since someone might suppose that the principles of the sciences are the same in that

way, i.e. due to [the fact that] the common axioms, which all sciences employ, are the same, for example that in every case there is either affirmation or negation, that things that are equal to the same are also equal to one another,[457] and things like that, for this reason he
316,1 says that not even in that way can the principles of all sciences be the same. For all sciences employ the common axioms, but of course they do not demonstrate the theorems in question through these [axioms] alone, but each science surely also assumes another proposition, in addition to the common axiom, from the genus that underlies it, and by combining this proposition with the common axiom it thus produces a demonstration.[458] For example, when [a science] wants to prove that the diagonal is incommensurable with the side, it assumes that it is commensurable, and through this [assumption] and the common axiom that says 'in every case there is either affirmation or negation', it demonstrates that it is incommensurable.[459] Likewise, assuming that this straight line is equal to that one and that yet another one is equal to that same one, [it] proves through that and the common axiom that <those>[460] straight lines are also equal to one another. And likewise in all cases. If, therefore, proofs not only go forward through the axioms, but it is surely necessary that from the underlying genus another premise is employed, which has the status of principle in relation to the conclusion, and the genera are not the same for all [sciences], it is pre-eminently clear that the principles are not the same for all sciences either.[461] And as has been said in the foregoing, neither are the axioms the same for all sciences,[462] nor is the manner in which the sciences always employ the common [axioms] <common> [to all sciences], but each [science] makes them appropriate to its own genus.[463] For example, 'whenever any four things are proportional, they will also be alternately proportional'[464] is a common axiom; and the geometer employs this axiom only in regard to magnitudes, saying 'whenever four magnitudes are in proportion', and the arithmetician 'whenever four numbers', and the natural scientist 'whenever four motions or four [intervals of] time'. So the sciences do not even employ the same axioms in precisely the same manner.

> **88b3** Moreover, the principles are not much fewer than the conclusions. For the premises are the principles, and the premises are made either of a term added by apposition[465] or of an inserted term.

Another argument. If, he says, the principles were the same for all sciences, they would also be finite in number. For otherwise the principles could not be known to be common to each science, if they were indefinite in number and, as it were, infinite. But now it is clear that those people who know [that the common notions are finite in

number] because it so happens that they are finite, and so declared [it to be their view] that every science uses the same principles, accordingly declared [it to be their view] that the principles are the same for all sciences. For just as of every word the principles are common and for this reason finite in number, I mean of course 24 letters, and the principles of every body are common, both the proximate ones, the four elements, and the first ones, matter and form,[466] likewise surely, even if the principles of the sciences were common, 317,1
they would have to be finite in number. As things are, there are many more problems than conclusions. For they are indefinite and infinite in number. And therefore the principles are infinite and indefinite in number as well. And because they are like that, they cannot be the same for all [sciences], as we have already said. Therefore the principles are not the same for all sciences. Now it is clear that the problems or conclusions are indefinite in number for each individual science and that to us they are infinite and boundless: for they will never be lacking. And that the premises, which are the principles of deductions, increase [in number] together with them, he proves in this manner: 'for the premises', he says, 'are generated either when a term is added by apposition or when a term is inserted'. For example, if the problem asked were 'for what reason is it the case that A belongs to B', then after a term has been inserted in the middle, two premises are generated, and one conclusion. For if C is posited as a middle term of 'A B', two premises are generated, namely AC and CB, and one conclusion. Again, if we wanted to prove the premise AC, by inserting as a middle term, for example F, then again two premises are added, and one conclusion. And thus always when a term is posited in the middle, there will be twice as many premises as there are conclusions.[467] So this is how [it goes], if the additional term is inserted in the middle. But if [the term] is added from outside, then when one term is added one premise and one conclusion are added, so that, when additional deductions are pieced together the premises outnumber the conclusions only by one. For example, if in the beginning there was a deduction that A belongs to B, and B to C, and therefore A to C, and subsequently a term D was added from outside, then one premise is added, and one conclusion: for B will belong to D through the middle term C. And if you want to infer that A will also belong to D, you will make a premise of the conclusion AC or BD. So the premises increase together with the conclusions and at the same rate, while the premises outnumber them only by one, due to the fact that, in accordance with the first deduction, when there are two premises surely one conclusion must be generated. If the premises relate to the conclusions in such a way that there are twice as many, (if the new term is inserted as a middle term) or one more (if it is added from outside), then in what sense does Aristotle say that 'the principles', i.e. the premises, 'are not much fewer than the conclu-

sions'? After all, it has been shown that they are more [numerous]. Well, I say that he was not concerned with giving precise information
318,1 about them, i.e. about how they are related, but, since it has been agreed upon that the conclusions are indefinite in number, i.e. [that they are] in that sense infinite, it sufficed to say only this much, that if they were not far fewer, then it would not be possible that the principles of the sciences are common. The premises could also be called fewer than the conclusions, if the same thing is not used twice, once as a premise, and once as a conclusion. And [the latter] is the case whenever a term is added from outside. For example, A [belongs] to every B, and B to every C, so A [belongs] to every C. If we add D, we add one premise, namely CD, but two conclusions, namely AD and BD. For if we say A [belongs] to B, and B to D, that B belongs to D has already been taken as a conclusion, and if we say that A [belongs] to C, and C to D, in like manner both conclusions have been taken.[468] But in this manner the result is that the premises generated are far fewer than the conclusions. For if more terms are adjoined, as many premises will be added as terms, but many more conclusions. For when a term is added, the [new] conclusions are as many as all those terms that lie outside, besides the pair.[469] For each of the higher [terms] will make a conclusion with the added one, apart from the penultimate, that is, the one before the added term; for that one makes a premise with the added term, not a conclusion.[470]

88b6 Again, the conclusions are infinite, but the terms are finite in number. Again, some principles are necessary, some can be otherwise. [Now if we consider them in this way, it is impossible for the principles, being limited in number, to be the same, while the conclusions are infinite in number.]

This is also [an argument] that establishes that it is impossible for there to be the same principles for different sciences. For some of the premises 'are necessary and others can be otherwise'. And from necessary things necessary [conclusions] are inferred, and from things that can be otherwise [conclusions] that can be otherwise. So it is impossible that the principles are the same for every science. For the principles of those sciences of which the conclusions can be otherwise will not be necessary, and neither will [the principles of those sciences] of which the [conclusions] are necessary be [principles] that can be otherwise.[471]

88b10 If someone means it in some different way, for example that *these* [are the principles] of geometry and *these* [are the principles] of calculations, etc. [...*these* are the principles of medicine, then what would be said other than that there are

> principles of the sciences? But it is ridiculous to say that they are the same, because they are the same as themselves. For in this way everything is the same.]

Again, he says, if 'someone were to say' that all the sciences have the same principles, not in the sense that everything is demonstrated from the same principles, but [in the sense that] there are different principles for each individual science, and that the theorems of medicine are demonstrated in one way, those of geometry in another way, and others in yet other ways, and [if] he were to say that the sciences have the same principles in the sense that 'they are the same as themselves', for example the principles of medicine are themselves the same as themselves, and likewise those of geometry and all the rest, then he would be saying something ridiculous, he says. For in this sense all things would be the same as one another. For each of them is itself[172] the same as itself.

> **88b15** However, neither [is it possible] that anything can be proved from everything[, that is, wanting the principles to be the same for everything. For this is too silly. For neither does this occur in evident teachings of mathematics, nor is it possible upon analysis. For the immediate premises are principles, and a different conclusion is generated when an additional immediate premise is taken.]

It is not possible in this sense either, he says, for the principles to be the same for everything, [namely] that from any given principle all things can be proved, and in general [that] each individual thing [can be proved] from all principles. For, he says, being false, it is refuted both on the basis of what is evident and on the basis of the argument. For neither do we see that in mathematics all things are proved from the same principle in this manner, not only the things in the different sciences (for example, the facts of geometry, and of medicine, or of music, are not proved from the same [principles]), but not even the [different facts] in one and the same science: for that the three angles of a triangle are equal to two right angles and that the diagonal is incommensurable with the side are proved from different principles. And likewise with the rest. So that, since we see that this occurs, saying that all things can be demonstrated from all [principles] is false from what is evident. And he also shows with an argument that it is impossible, in this manner: when we analyse the deductions, he says, we arrive at the immediate premises. For we analyse up to that point, until we arrive either at definitions, or at immediate premises. Now neither the definitions nor the immediate premises are the same for all things. For after another immediate premise has been added, he says, another conclusion results as well. For example, 'Socrates is

a substance'. When we analyse through which premises that has been inferred, for example that he is a human being, and that [i.e. human being] is an animal, and so forth, when we arrive at the immediate premise that body is a substance, we stop the analysis. Likewise, when we analyse 'the number two is a quantity', we say that it is even, and that this [even] is a number, and that [number] is discrete, and that [the discrete] is a quantity. The premise that says 'the discrete is a quantity' is a different immediate premise from the one that says that the body is a substance. The same holds of definitions: for when we analyse 'the three angles of a triangle are equal to two right angles', we arrive at the definitions of straight line and angle. And likewise with the other things. And the definitions
320,1 are different, because the things are different as well. 'In the evident parts of mathematics': that is, in those parts of mathematics that are clear to and generally agreed upon by us.

> **88b20** If someone were to say of the primary immediate premises that they are the principles, then there is one in every genus.

[He means] that not even in this way is it possible that the principles are the same, by the immediate premises, from which deductions [are made], being the same. For there is one immediate premise, he says, in every genus, for example in arithmetic that the unit is indivisible, in geometry [that] the point [is indivisible], and other [premises] in other genera. So since the genera are different, the immediate premises also differ. What someone would say that that one and first immediate premise[473] [is] in music, or medicine, or physics, or the other sciences, I cannot say. But that cannot be seen in the ten categories, either. After all, for each of them, the first immediate [premises] are at least two, if indeed they are divided into the kinds that are opposed to each other – substance into body and the incorporeal, quantity into continuous and discrete – and the genus is predicated immediately of each of the proximate members of the division.[474]

> **88b21** And if [he says] neither that anything must be proved from all of them, nor that they are different in the sense that they are different for each science, [[the question] remains, if the principles of all things are akin, while different things are [proved] from different [principles]. And it is clear that this is also not possible. For it has been shown that the principles of things that are different in genus are also different in genus. For principles are of two kinds, those from which, and those about which[475] [we prove]. Those from which [we prove] are

> common, and those about which [we prove] are specific, such as number and magnitude.]

Going through all the ways in which we saw that it is possible for the principles to be the same for all sciences, he concludes what has been said, and adds the as yet remaining possibility. For if 'neither anything is proved from all of them' (and that can be understood in the two ways [discussed], either that each [thing] is proved from every principle, or that all things [are proved] by each) – if, then, neither are the principles for all things demonstrated the same in this way, nor is it possible for that silly assumption to hold, that because each science has its proper principles, each is itself the same as itself, then one more way, he says, is left over, according to which it is possible to say that the principles are the same. For maybe someone will say that the principles are the same in this way, not that everything is proved through the same [principles], but that they are all akin [i.e. of the same genus], some being useful for this science, some for that, just as the same geometrical principles are all [principles] of geometry, but different [principles] are useful for different theorems: for perhaps the point, the line, the plane and the rest are principles, but through [the proposition] that the point is partless this theorem is perhaps demonstrated, but through the line another, and yet another
through something else.[476] So what prevents the principles of all 321,1
things there are from being akin, although different things are proved through different principles, facts of geometry through some, facts of arithmetic, or physics, or whatever else, through others? Because, he says, it is not possible in this way either, and this is clear from what has been demonstrated before. For it has been shown that the principles are akin to the things that follow from them. And things are different in kind [i.e. genus], so the principles are different in kind [i.e. genus] as well. For since the principles are of two kinds, as has been shown, namely [those] 'from which' the demonstrations are [made] (these are the common notions) and [that] which the demonstrations are 'about' (these are the immediate premises taken from the genera underlying the individual sciences), the premises 'from which' the demonstrations [are made] are the same (I do not mean that all the common notions are of the same genus, but that all the sciences use them[477]), but as for the premises 'about which', those are necessarily different in genus. For number is not the same as magnitude, for they are opposed [to one another]. Magnitude is continuous quantity, and number is discrete quantity. So it is clear that the premises employed in the case of number, too, would be completely different from the ones employed in the case of magnitude.

It is in no way possible, then, for the principles to be the same for everything.

Chapter 33

> **88b30** Scientific understanding and its object differ from opinion and its object, etc. [...because scientific understanding is universal and through necessary things, and the necessary cannot be different.]

This theorem too is appropriate to the accounts about scientific understanding, because it allows us to distinguish opinion from scientific understanding and the object of opinion from the object of scientific understanding. It distinguishes opinion from scientific understanding by the mode of apprehension,[478] and the object of opinion from the object of scientific understanding by showing that the objects underlying opinion and scientific understanding are different. For distinguishing the activities from one another does not suffice at all for showing that the things the activities concern are different as well. For he next raises the question[479] whether perhaps, even if opinion and scientific understanding are different, still the object might be the same,[480] so the very same thing which is the object of opinion, can also be the object of scientific understanding, so there would be both opinion and scientific understanding of the same thing. But in that case, again, how could opinion and scientific understanding not be the same?[481] So for this reason he shows both, i.e. both that opinion is something other than scientific understanding and that the object of opinion [is something other] than that of scientific understanding. By opinion he means true [opinion]: that is what he wants to distinguish scientific understanding from, because false opinion in no respect resembles scientific understanding.

Now scientific understanding differs from opinion, in that scientific understanding is conviction that cannot be persuaded otherwise, and always stays the same, but opinion is conviction that can also be otherwise. From these [characteristics] it is also clear with respect to which things they are active, I mean [what is] the object of scientific understanding and that of opinion: for the object of scientific understanding is what is by necessity and could not possibly be otherwise, such as that fire warms or that the heaven moves in a circle;[482] the object of opinion, however, is that which by nature is sometimes this way, and sometimes that. Moreover, opinion and scientific understanding are different in another way, if indeed one concerns one object and the other another, so that scientific understanding differs from opinion in these two [respects], namely by the mode of apprehension and by the object the apprehension concerns.[483] Evidence that opinion is concerned with things that can be otherwise and that we are accustomed to call the understanding we have of those things opinion, comes both from the argument and from people's common usage [of the term]. For of the things that are,[484] he says,[485] some are such that they can never be otherwise, and some just are the case, but not from necessity: instead they can also not be the case. And it is clear that when we make a division of the things that are we will not include the things that are impossible: for those are not things that are, such as that man is irrational. Now of which kind of things, he says, would opinion be the apprehension?[486] Perhaps of those that allow both of being the case, and of not being the case, such as that Socrates is bathing or not, or that he is philosophising or is not philosophising. It is clear that scientific understanding is not concerned with such things; for it is a kind of standstill,[487] and only of things that are judged as not such as can be different – and from this it took its name;[488] opinion on the other hand is simple apprehension,[489] that can also alter. For it is not impossible that someone who has the opinion that the soul is immortal changes his mind again upon encountering persuasive arguments that it is mortal. But of course someone who has scientific understanding that it is immortal cannot possibly ever believe that it is mortal. Now if it is also possible for the things that can be otherwise to be different, and we have scientific understanding about those things, but scientific understanding concerns things that cannot be different, then the things that can be different could not possibly be different,[490] which is impossible and absurd. If therefore it is not possible that scientific understanding concerns these things, it follows that opinion is concerned with them, quod erat demonstrandum.[491] And people's common usage confirms it.

When we know something for certain, we are accustomed to use the [expression] 'I understand',[492] but whenever we [believe] something that can also be otherwise, we say 'I think' and we say that this is how most people think about this, while we never use this kind of word in the case of things that are by necessity – instead, we use 'we understand' or 'we know' or 'we have knowledge' or something like that.[493]

From what has been said it is clear that in this passage he calls opinion, not the knowledge of the fact without the reason why,[494] but, as I said, the apprehension concerning the possible. The apprehension concerning necessary things, if it should be without deduction and explanation, he calls scientific understanding of the fact, and if with deduction and explanation, scientific understanding of the reason why. Likewise, opinion without deduction is only of the fact, and [opinion] with deduction is both of the fact and of the reason why.[495]

88b32 There are things that are true and are, but can also be different. [So it is clear that there is no scientific understanding about these things. For then it would be impossible for the things that can be different to be different.]

After saying in what [kind of things we find] scientific understanding, he also wants to say in what [kind of things we find] opinion. After saying that 'there are things that are true' he added 'and are'. And Alexander, in explaining the passage, says that he added 'and are' for this reason, that after all there is truth in the case of what is not, too, such as when I say that a goatstag is not[, i.e. does not exist].[496] The philosopher[497] said that Alexander was not right in saying this. For, he says, it does not belong to opinion to know such things but to scientific understanding. For that which is not, and which is truly said not to be, cannot possibly be different. And that [kind of] apprehension is not a matter of opinion but of scientific understanding. So by parallel reasoning he said that we should understand 'true' and 'are' as having the same extension. But since there is both the necessary, or rather the impossible, also in the case of things that are not, but there is that which can be otherwise as well (for whenever I say 'a goatstag is not', I said something that is true and it is impossible [for it] to be, but when I say 'now I do not bathe, now I do not walk', while indeed neither bathing nor walking, I just said something that is true, but it can nevertheless be otherwise), then it is clear that of that which is not, but contingently so, there would be opinion, not scientific understanding, so that the interpretation of Alexander would hold ground.[498] For 'are' would be added with a view to excluding[499] the things that are true and can be different, but are not [the case], such as when someone who is not bathing says he is not bathing.

> **88b35** But then [there is] no intellect [of these things] either, for I call intellect the principle of scientific understanding, nor [is there] indemonstrable scientific understanding, for that is the apprehension of an immediate premise.

Because he wants to show that of the soul's rational capacities only opinion is concerned with things that can also be different, he presents us with a division of the soul's rational capacities. And he calls one of them intellect, one scientific understanding, and one opinion. He divides scientific understanding in two, namely on the one hand knowledge through deduction, of things that are necessary and always the same, and on the other hand indemonstrable scientific understanding. The indemonstrable [kind] he also divided earlier into the apprehension of immediate premises and the knowledge of common notions, from which, together with the immediate premises, demonstrative deductions are generated. And he calls this, I mean the indemonstrable [kind of knowledge], scientific understanding of the fact without the reason why. In a similar manner he also divides opinion, as we have already said above. And intellect he calls the highest capacity of the soul, by virtue of which there is apprehension of the divine, in accordance with the so-called simple intuitions.[500] He also said in the foregoing that 'we say that there is not only a kind of scientific understanding, but also a principle of scientific understanding, by which we know the definitions',[501] where he calls the divine and intelligible Forms 'definitions', as definitional and, as it were, the limits of beings. He calls that kind of intellect 'principle of scientific understanding', because the grasp of common notions and generally speaking the knowledge of immediate premises that comes to us without deduction is a kind of last and lowly activity of that intellect.[502] It is clear, then, that in accordance with what is taught here 'thought' would be predicated jointly of both scientific understanding and opinion [generated] through deduction, since it [sc. thought] is the name of a certain kind of activity of theirs.[503] Enough, then, about the soul's capacities. Having shown that scientific understanding [generated] through deduction does not concern what can be otherwise if indeed it proceeds through necessary things and things that are always the same, whereas what can be otherwise is different at different times, he shows that a fortiori neither will intellect concern what can be otherwise, if indeed it is the principle of scientific understanding. And surely neither will indemonstrable scientific understanding, for that too is a beginning of scientific understanding [generated] through deduction. What remains, then, is that opinion concerns things that can also be different – and it differs from scientific understanding in just this respect. And proving that was just the [issue] before him.

> **88b37** Now intellect, scientific understanding, and opinion, and what is said through them, are true. [So it remains that about the true or false, which can also be otherwise, [we have] opinion.]

This is as it were the conclusion of what has been said. For if the capacities of the soul by which we speak the truth are just those mentioned, I mean 'intellect and scientific understanding and opinion', and the things that 'are said through them', that is, all things proved by means of these [capacities] through a deduction, in order to specify the scientific understanding and opinion we have through a deduction, and [if] of the things that are, some are necessary, namely just the ones that are and cannot be different, and others can be different, and scientific understanding concerns the necessary things, then it remains that opinion concerns the things that are true – not the necessary ones, though, but the ones that can also be different. And that would be the difference between them.

> **89a3** That is, the apprehension of an immediate and not necessary premise. 325,1

That is, opinion, which concerns things that can be otherwise, is 'the conviction of the immediate but not necessary premise'. For we have also said before that there are not only demonstrative immediate premises, but also dialectical ones and reputable ones. For if by virtue of opinion[504] it is agreed that pleasure is a natural activity, then such a [proposition] is a dialectical immediate premise. And likewise, that according to Plato the soul is immortal: that is likewise an immediate premise, and a reputable one.

> **89a4** And this agrees with the appearances, because opinion is something uncertain, and nature is like that as well. [Moreover, no one thinks he has an opinion, when he thinks it is impossible [for something] to be different. Instead, he thinks he has scientific understanding.[505] But when [he thinks] that it is like this, but nevertheless nothing prevents it from being different, then he [thinks that he] has an opinion, assuming that of such a thing there is opinion, but of the necessary there is scientific understanding.]

He gives evidence that opinion concerns things that can be otherwise on the basis of common usage as well. For we say that opinion is 'uncertain', due to the fact that its object can also be different. And when they alter, the opinion about them alters as well, either by becoming false instead of true, or by changing completely.[506] So this is what we mean when we say that the opinions of humans change along with the facts. But one would not say such a thing in the case of scientific understanding, that the states of scientific understanding change or that scientific understanding in general is something uncertain. And then, he says, 'not one' human 'thinks that he has an opinion' about something, when he knows that that thing cannot 'be different'. 'Instead, he thinks he has scientific understanding'. And [he thinks] 'that he has an opinion', when he thinks that [the thing in question] can also be different. For who would ever say 'I have the opinion concerning the sun that it will rise tomorrow'? Instead, [he would say that he has the opinion] that there will be rain, for example, or war, or something like that, precisely because he thinks opinion concerns what can be otherwise, but scientific understanding concerns what is necessary.

> **89a11** So in what sense is having an opinion not[507] the same as having scientific understanding, and why will opinion not be scientific understanding, if one were to posit that one can believe[508] everything one knows?

With these [words] he sets out a certain puzzle which we have already spoken of before. For can it be, he says, if it is possible for the same thing at the same time to be the object both of opinion and of scientific understanding, that opinion and scientific understanding are the same [thing] as well? For example, that there will be an eclipse, if you will, after so many days. If someone were to think that, in the sense that it cannot be different, but that this will instead happen by necessity, then this [person] has scientific understanding of it. If, on the other hand, someone were to think about that same thing that it will be, but not at all that it will happen by necessity, but that it could also not happen, then he has an opinion concerning the very same thing, but not scientific understanding. Likewise, if someone were to think that for a triangle it is also possible sometimes not to have its three angles equal to two right angles, he has an opinion about that, not scientific understanding. But if someone were to think that it can never be different, [he has] scientific understanding. So one can have an opinion and scientific understanding of the same thing, and the object of opinion is not distinguished from that of scientific understanding. This is clear, he says, also from this: just as [it is possible] to have an opinion and scientific understanding about the same thing without a deduction in the aforesaid manner, [i.e.] knowing only the fact either in the manner belonging to opinion or in that belonging to scientific understanding, it is also possible [sc. to have an opinion and scientific understanding of the same thing] through a deduction, if one person establishes that the triangle has three angles equal to two right angles through necessary [premises], and the other [does it] through premises that can be otherwise, and each of them analyses it up to the point where he reaches the immediate premises – the first one [reaching] necessary premises and the second one [reaching] premises that can be otherwise. So if indeed it is possible to deduce the same thing both in the manner belonging to opinion and in that belonging to scientific understanding, and in each case the analysis [reaches] the immediate premises, then it seems that it is possible for there to be both opinion and scientific understanding of the same thing, and the object of opinion does not differ from that of scientific understanding. And if this is so, then opinion and scientific knowledge are the same as well. Enough about the puzzle. After first having worked through the text of the puzzle, we will consider its solution afterwards.

'So in what sense is having an opinion not the same as having scientific understanding, and why will opinion not be scientific understanding, if one were to posit that one can believe everything one knows?' The puzzle stops here. If someone were to suppose, he says, that of that of which there is scientific understanding, one can also have an opinion in the way I have said then how would the same thing not be an object both of opinion and at the same time of scientific understanding?[509]

> **89a13** For the one who knows and the one who has an opinion will follow [the same thread] through the middles, until he comes to the immediates.

'He will follow [the same thread]', instead of 'he will analyse the deduction',[510] 'through the middle' terms until 'he comes to the immediate' premises. For example, it is possible to establish through premises that can be otherwise that souls are immortal, in this manner: all men honour the graves of their ancestors; those who honour the graves of their ancestors do that for the sake of tending to the departed; those who do this for the sake of tendance believe that the ones they tend to, exist; for nobody would tend to someone who is not there; therefore, the souls of the departed exist and did not perish together with the bodies; [souls] which, when the body perishes, do not perish with it, but are [still] there after its perishing, would be immortal; therefore the souls of men are immortal. It is possible for someone who analyses the deduction, to analyse it up to this point, until we reach the immediate premise, I mean of course that all people honour the graves of their ancestors. For this is an immediate premise of opinion that has credibility from what is evident.

> **89a14** So that if indeed the former knows, then the person who has an opinion knows as well. [For just as it is also possible to have an opinion of the fact, so too [is it possible to have an opinion] of the reason why. That is the middle term. Or, if he has assumed the things which cannot be different in the way [he has assumed] the definitions through which demonstrations [are made], then he will not have an opinion, but scientific understanding. If [he has assumed that] they are true, but not that these belong to those on account of essence and of form, he will have an opinion and he will not truly have scientific understanding, both of the fact and of the reason why…]

That is, just as he who has scientific understanding, by ascending to the immediate [premises], knows not only the fact but also the reason why, so he who has an opinion [knows them] as well, in virtue of the same cause. That this is what he means, is clear from the things he goes on to say. For, he says, 'just as the fact' can be known the way opinion knows, so can 'the reason why'. For the conjunction 'for' is causal. For if the middle term through which the deduction [is made] can be otherwise, then the opinion is of the fact; if, however, [the middle term] is necessary, then [the opinion] is of the reason why. So just as there is also scientific understanding of the fact and of the reason why, so there is opinion [of the fact and the reason why] as well. So if it is possible to know the same thing, now through necessary middle terms, now through ones that can be otherwise, then one can have both an opinion and scientific understanding of the same thing, quod erat demonstrandum.

> **89a21** ...if he has an opinion through immediates. [But if he does not [have an opinion] through immediates, he will only have an opinion of the fact.[511]]

We should supply in thought 'of the reason why', i.e. [if he has an opinion through immediates] he will have an opinion of the reason why. For after saying 'he will have an opinion and he will not truly have scientific understanding both of the fact and of the reason why' he goes on to say the following in addition to the adjacent [phrase] 'why': 'if he has an opinion through immediates', that is to say, [an opinion] of the reason why.[512]

> **89a23** There is opinion and scientific understanding of the same thing, but not [in every sense]. [Instead, just as there is true and false opinion of the same thing in some way, so there is scientific understanding and opinion of the same thing as well [i.e., in some way]. For that there is a true and false opinion of the same thing in the way some say, results in having all kinds of absurdities, including that someone does not have an opinion of that of which he has a false opinion.]

After having raised the question, how scientific understanding and opinion would not be the same, since there can be opinion and scientific understanding of the same thing, he now, in this [passage], adds the argument and the solution to the puzzle. And he says that there can be opinion and scientific understanding of the same thing, in the same way in which there can also be a false and a true opinion of the same thing. Having articulated earlier[513] how there can be a false and a true opinion of the same thing, he thus shows on the basis of that example how there can be both scientific understanding and opinion of the same thing. Now there is, he says, false and true opinion of the same thing 'in some way' – namely to the extent that there is also false opinion about that of which there is true opinion.[514] For example, if one [person] has the opinion that the soul is immortal, and the other that it is mortal: for by there being both a true opinion and a false one concerning one and the same object, there is a true and a false opinion about the same thing in this way, but not in the sense that the true and the false one are identical to one another. For a number of absurd things would follow from [the latter] argument. First, that the true [opinion] would be the same as the false one, and that speaking the truth would be the same thing as a falsehood. That means that [the members of] a contradiction are simultaneously

10 true. For contradiction divides true and false. Then, he says, it will
follow that someone will 'not have the opinion that he has'. For if we
were to have the opinion that, for example, the soul is immortal, and
that is true, then it is clearly false that [the soul] is not immortal.
Now if the truth were identical to that,[515] then having the opinion
that the soul is immortal, <would be the same as not being of the
opinion that it is immortal>,[516] which is absurd. Now as it is in the
case of true and false opinion, so [it is] concerning opinion and
15 scientific understanding as well. For because there is both opinion
and scientific understanding of the same object, both are of the same
[thing]. For both he who has scientific understanding and he who has
an opinion will say that the three angles of the triangle are equal to
two right angles, and both these apprehensions are true, and in that
way there is opinion and scientific understanding of the same thing.
They differ in manner, because scientific understanding says of what
20 holds by necessity that it holds by necessity, whereas opinion [says
that it holds] contingently. For example, he who has scientific under-
standing will say that the sun is in motion by necessity, and he who
has an opinion [will say] that it is in motion contingently. So they
differ in the manner of their apprehension. So it is possible for both
to be about the same thing: but it is impossible [for them] to be the
same, just as also the false opinion is not the same as the true [one].
For saying that that which holds necessarily holds contingently is
25 also false. So they have the object in common, by both being about
one and the same thing and [by both] being true about that [thing],
but they differ, because scientific understanding is also true in the
mode of predication, while opinion is false [in that respect]. And if the
mode of predication is a different one, it is clear that the object of
30 scientific understanding cannot be the object of opinion either. For
that the soul is immortal is truly [taken to be] an object of scientific
understanding, but falsely [taken to be] an object of opinion. For if
that which is the object of opinion can also be different, and [if] [the
fact] that the soul is immortal, if it is immortal, cannot be different,
and [the fact] that a triangle has its three angles equal to two right
angles cannot be different either, then that which is the object of
scientific understanding cannot be the object of opinion. Now an
329,1 opinion about [such a fact] can be true, but [such a fact] cannot be the
proper object of opinion, since scientific understanding cannot be
opinion either.

89a28 Since 'the same' is said in more than one way, there can be [opinion and scientific understanding of the same thing] in one way, but not in another.

5 Now if that which is said to be <the>[517] same as something were the
same in all respects, there cannot be opinion and scientific under-

standing of the same thing. But since '"the same" is said in many ways' (for things are said to be the same in subject or definition or in whatever other way), nothing prevents that there is both opinion and scientific understanding of the same thing in some respect, and not in another, as we have already said.

> **89a29** For having the true opinion that the diagonal is commensurable is absurd. [Instead, they [i.e. the true and false opinion] are [opinions] of the same in this sense, that the diagonal, which the opinions concern, is the same.]

He also develops the argument with respect to an example. For, he says, saying that the true and the false opinion are of the same thing in the sense that they are the same as one another is absurd, like the [opinion] that says that <the diagonal is incommensurable and the one that says that> it is commensurable.[518] For the true will [then] be the same as the false. So in what sense is it that the true opinion and the false one are [opinions] of the same thing? By being about one and the same object, the diagonal.

> **89a32** But the essence of each [of them] according to its definition is not the same. [In the same way [there can be] scientific understanding and opinion of the same thing.]

For it is clear that the definitions of true and false opinion are different and also [the definitions] of opinion and scientific understanding.

> **89a33** For the one [i.e. scientific understanding] is of animal in such a way that it cannot [not be an animal, but the other [i.e. opinion] [in such a way] that it can.]

Scientific understanding, he says, will declare [its view] concerning 'human being' that it is an animal, in the sense that it is impossible for it not to be an animal. But opinion [will declare it in the sense] that it is contingently an animal.

> **89a35** For example, if the one [i.e. scientific understanding] is of precisely what is human being, but the other [i.e. opinion] of human being, but not of precisely what is human being. That it is human being is the same, but the manner is not the same.[519]

That is, scientific understanding and opinion about Socrates, for example, suppose that he is a human being, and in this respect both are true. But scientific understanding says that Socrates is precisely what is a human being, and that is the same as saying that he is not not a human being, and that [means] that it is not possible [for him] not to be a human being, but opinion [says] that it is also possible [for him] not to be a human being. In this respect, then, there is a difference, according to the manner of apprehension.

> **89a38** It is clear from these things that it is not possible to have an opinion and have scientific understanding about the same thing at the same time. For [the person in question] would simultaneously have the apprehension that [the same thing] can and cannot be otherwise. [And this is not possible. For it is possible for each [of them] to be of the same thing in a different [person] in the way we have said, but in the same [person] it is not even possible in that way. For he will have, at the same time, both the understanding, for example, that human being is precisely what is animal (for that, we saw, is that it is not possible not to be animal) and [the understanding] that [human being] is not precisely what is animal – let that be that it is possible [not to be animal].]

If it has been shown that it is possible that there is opinion and scientific understanding of the same thing in some, and that it is not possible in some respect, and that it is not possible in the sense that it is impossible for scientific understanding and opinion to be the same, it is of course clear from this that it is also not possible to have an opinion and scientific understanding about the same thing at the same time. And he rightly added 'at the same time'. For it *is* possible [to have an opinion and scientific understanding] at different times. For it is possible for people who first have an opinion that the sun is eclipsed because the moon passes in front of it [supposing] that this does not happen by necessity as they say Epicurus believed – to later have a scientific apprehension about it, because they have talked to astronomers.[520] But really, for one and the same person to have both an opinion and scientific understanding that this occurs like that, at one and the same time is something impossible. For he will suppose the same thing at one and the same time to both be able to be otherwise and not be able to be otherwise, which is absurd. For insofar as he has scientific understanding of it, he supposes that it is not possible for it to be otherwise. But insofar as he has an opinion, he supposes that it is possible for it to be otherwise as well.

331,1 **89b7** How we should distribute the rest among thought, intellect, scientific understanding, craft, prudence and wisdom…

Since throughout the book he has elaborated separately on demonstration and opinion, but there are also other capacities, or rather activities, of the soul, which he himself enumerated, he wanted to tell us the reason why he did not elaborate separately on them in this work. And he says that elaborating about those things separately does not belong to the present work, but [it belongs to the work on] ethics [to elaborate] about some of them, and to [the work on] physics [to elaborate] about others. We would add that for some [it belongs to the work on] theology [to elaborate on them]. At all events he has discussed prudence and craft in the work on ethics;[521] for both craft and prudence concern things that are done. In the work on theology, I mean the *Metaphysics*, and in *Alpha Elatton*,[522] [he discussed] intellect and wisdom, and scientific understanding and thought and opinion he discussed in the logical and physical works.[523] But perhaps we may as well say a little about them right now. Now prudence and craft concern things that are done, things that can be otherwise and what is up to us; this is why it is no surprise that he discusses them in the ethical work as well. Now concerning craft both its end and the means to the end are distinguished. For the craftsman does not give forms to the products of his craft as he pleases, but as the intended use assigned [them]. For it is not up to a craftsman to give whichever form he wants to a boat, or a bed, or a house, but [to do it] in the way determined by the intended use. Prudence, however, both has an indeterminate end and employs sometimes one means to the end, sometimes another. For example, we decided to wage war against those [people], or to depart to that place, and the plan was not realised entirely because of either death, sickness, or a storm, or several other things. Also, the means to the end are not the same: for perhaps it is possible [to reach the end] both if one goes on foot, and if one uses beasts of burden or boats. And often we planned to use boats, but this [plan] was not realised. All these things can both be otherwise and are up to us. Opinion is of wider extent than both craft and prudence. For it even extends to the things that are not up to us. For example, when we behold the moon after conjunction [i.e. when it is waxing[524]], and it has blunt horns, we say that therefore [there will be] rain, and all the things [people] write about the weather signs in the sky. These things can be otherwise, but are not up to us. Thought is more extensive than opinion, because it is said to occur not only in the case of deductions of opinion or dialectical [deductions], but also in the case of demonstrative [deductions]. It has often been said that everyone uses thought, [which is] the same as [saying that everyone uses] the deductive procedures, including craftsmen and actions.[525] We mentioned that scientific understanding is said to occur not only in the case of demonstrations with a deduction, but also in cases without deduction, I mean of course the apprehension of the immediate premises. So in that respect it is of wider extent than thought. But insofar as thought is also [referred to] concerning the objects of opinion, in that respect scientific understanding is of smaller extent than thought. Intellect and wisdom[526] are either entirely the same, or, if one wanted to divide them more precisely, the summit of intellect and as it were its perfection, that is what is called wisdom,[527] in accordance with which there is apprehension of the divine and intelligible forms, and it is called *sophia* as if it were a kind of clarity (*saphia*), because through it divine things become clear to us.[528] For divine things, as Aristotle himself says, are most apparent and most clear; most apparent by their nature (for they are pure forms, without matter, beings in act without potentiality), and most clear, because our knowledge is generated in accordance with the forms. For this reason matter cannot be known by a rational principle proper to it [i.e. to matter], as it is formless, but can be grasped by a bastard reasoning, as Plato says, because we cannot grasp its nature by a direct apprehension, because it does not have any form.[529] So all things that are pure form without matter, are clearly most apparent and most clear. And they become unclear to us not because of themselves, but because of our weakness. Just as also the sun, which is most apparent and best known, more than all the other stars and all other beings, is invisible and obscure to bats, while the other stars are easier to distinguish, due to the weakness of their power of sight, just so are divine things [obscure] to us.[530] So since wisdom, which comes to our souls when they have been perfected, provides the apprehension of the divine, for this very reason, which I mentioned, it has been allotted that name,[531] clarifying to us, as it were, what first seemed to be unclear and unknowable. So that is the task of wisdom. [The task] of intellect is the apprehension of definitions and of immediate premises. And for this reason [Aristotle] said that intellect is 'the principle of scientific understanding', 'by which we know the terms'.[532] Now the discovery of the ten categories, which are the principles of all scientific understanding and of all deduction, should be ascribed to nothing other than to the intellect.[533]

89b9 Some belong rather to physical, others to ethical theory.

By 'physical' he does not mean the theory that discusses physical
333,1 things, which we specifically call philosophy of nature, but in a general sense that which discusses beings insofar as they are beings. That would be either physical [beings] or those which transcend them.

Chapter 34

89b10 Acumen is a certain capacity of hitting upon the middle term in an imperceptible time, [for example if someone, on seeing that the moon always has its luminous side towards the sun, quickly thought of why this is, [namely] because it shines [with light] from the sun. Or he knows that someone is talking to a rich man, because he is borrowing money from the latter.

Or [he knows] why they are friends: because they are enemies of the same [person].]

In the cases enumerated he investigates what acumen might be. And he says that it is 'hitting upon the middle term in an imperceptible time', that is, the discovery of the cause of the problem, which becomes the middle term in the problem and makes a deduction. 'In an imperceptible time' instead of 'in a brief time',[534] in order for the discovery of the middle term to occur without thinking about it. And the examples are clear. For example, he says,[535] if someone upon having been asked why the moon 'always has' its illuminated side 'towards the sun', replies at once and without thinking that 'it receives its luminosity from the sun', clearly not knowing this before, but instead now having discovered it, such a person is called acute, and such an activity is acumen.[536] Likewise, if someone, upon seeing a poor man talking to a rich man, spots at once that he wants to borrow money,[537] this person likewise is acute. Likewise, if someone, upon seeing two friends who have been initiated together,[538] spots at once that they are 'enemies of the same [person]', or that they love the same [person], or something like that, then such a person is acute. For really, such equivocal friendships of the masses do not in any way bring along true friendship. For friendship is a kind of divine and unifying good, which, as Plato says, transcends justice, and whereas justice needs friendship, friendship does not need justice.[539] But the masses pursue the thing falsely called [friendship], and are enemies really, rather than friends. And it is this kind of friendship of the many that Aristotle here speaks of.[540]

89b14 For upon seeing the extremes he[541] recognised all the middle terms that provide explanations. [A stands for 'the side towards the sun being luminous', B for 'shining [with light] from the sun' and C for 'the moon'. B, shining [with light] from the sun, belongs to the moon, C; to B belongs A, that the luminous side is [directed] towards that, [with the light] from which it shines; so A also belongs to C, through B.]

The acute [person], he says, 'upon seeing the extremes' at once recognised 'all the middle terms', which are both explanatory and what connects the extremes. And he subsequently gives an outline,[542] of what are the extremes and what the middle term, which [the person in the example] recognised. What he says is clear and it has already been said by us.

Notes

1. Aristotle begins the lemma by considering *sullogismoi* in general; within the next few lines he makes clear that here, as elsewhere in the *An. Post.*, his special concern is the sort of *sullogismos* called a demonstration. Within the *An. Post.*, it is often unclear whether *sullogismos* has its technical sense, as referring to a deduction that meets the formal requirements laid out in the syllogistic theory of *An. Pr.* 1, or whether it has its looser sense, according to which it refers to any deduction whatever. This has led some interpreters to identify different strata of the *An. Post.*, one, prior to Aristotle's formal syllogistic theory of *An. Pr.*, which does not require demonstrations to conform to any particular formal structure, and a later one, including the present chapters, which takes demonstrations to be canonical Aristotelian syllogisms; see Solmsen (1929); Barnes (1981). Philoponus himself, like all of the ancient commentators, is oblivious to diachronic considerations in his interpretation of Aristotle; throughout, he tries to make sense of Aristotle's work as a consistent whole. Nonetheless, because his reading of Aristotle's text is so close, he is not wont to import Aristotle's theory of the syllogism into passages where Aristotle himself does not appeal to it. For this reason, we translate *sullogismos* as 'deduction' except in those cases in which there is explicit appeal to the syllogistic theory.

2. The term 'theorem' here has its technical sense, a proposition to be proven. See Proclus, *in Euclid* 77-81.

3. Philoponus, following his teacher Ammonius, referred to below as 'the Philosopher', takes Aristotle to be raising a question not concerning all syllogisms, in general, but only demonstrative syllogisms: can they be extended indefinitely, by means of taking the major term of the original demonstration, and employing it as a middle term for a new demonstration?

4. This would be true only if the predications that can form a sequence that proceeds to infinity are of the sort that can serve as demonstrative premises. Aristotle would, however, have to sanction at least some predications of a sort that can be extended to infinity. Since every magnitude is such as could be bisected, and the process is potentially infinite, through such a process of bisection there could be a chain as follows: length AB is twice the length of AB; what is twice the length of AB is four times the length of AC, what is four times the length of AC is eight times the length of AD, etc. Analogous chains could likewise be generated for an increasing series of numbers (a unit is half of two, what is half of two is a quarter of four, etc.). For the results of these chapters to stand, Aristotle, on Philoponus' interpretation, would have to somehow deny that these predications could serve as demonstrative premises.

5. The present commentary is presented as a record by Philoponus of the classes given on *An. Post.* by Ammonius, with additional remarks by Philoponus himself (37,1-4). Here, as elsewhere (as at 47,24), Ammonius is referred to as 'the Philosopher'.

6. An immediate premise is one that cannot itself be demonstrated; that is to say, within a demonstration there can be no middle term between its subject and predicate, which would result in two premises that are 'prior' in the order of demonstration. See *An. Post.* 1.2, 72a8. Aristotle had argued at 72b7-72b14 that there must be immediate premises on the grounds that there can be no demonstrations of infinite length. He there argued against the possibility of infinitely long demonstrations on the presumptions that scientific understanding does indeed result from demonstration, and that it is impossible to mentally traverse an infinitude. (The remainder of 1.3 dispenses with the possibility of infinitely cycling demonstrative chains consisting of a finite number of terms.) Philoponus, following Ammonius, is suggesting that Aristotle is supplementing this earlier argument. He can be understood as doing so in two ways. First, Aristotle's earlier argument for the impossibility of infinite demonstrative chains rested on the assumption that any given demonstration arises from a finite number of immediate predications. He left untouched the possibility of demonstrations being indefinitely extended by adding terms outside the minor or middle terms. This possibility is the concern of the present argument. Second, Aristotle's earlier argument rested on an epistemological premise: that scientific understanding, from demonstration, does exist. The arguments from 82a3 on deal with the nature of predication in general, for which reason, Philoponus reports, Ammonius thinks they are termed 'formal'. See n. 77.

7. i.e. Aristotle.

8. *An. Post.* 1.15, 79b13-20.

9. This is a provisional conclusion. The previous argument did not show that infinite predicative chains are in all cases incompatible with the nonexistence of immediate predicates. But, as Philoponus proceeds to point out, the two can indeed be shown to be incompatible.

10. As Mignucci (1975), 394 points out, while Aristotle is presumably here concerned with denying the impossibility of infinite predicate chains of the sort that might feature in demonstrations, Philoponus takes Aristotle to be here denying the possibility of any infinite syllogistic chains at all, a result that he takes to be preliminary to proving the impossibility of infinite demonstrative chains.

11. Philoponus at 35,4 takes Aristotle at 72a18-20 to mean that any unmediated predication is the sort of principle we call a hypothesis. See McKirahan (2008), 126 n. 199. He is, of course, not here saying that any two such hypotheses yield a demonstrative conclusion; they need to stand in the appropriate relation to each other.

12. Instead of *hôste ei kai esti têi alêtheiai tôn AB meson, dokei de mê*, Ross has *hôste ei kai mê esti ti têi alêtheiai tôn AB meson, dokei de einai*, which could be translated 'So even if in truth there is not a middle term for AB, but there seems to be one.'

13. As Ross (1949), 568 points out, *toutou* at 81b22 seems to modify an implicit *mesou*, which would support the alternative manuscript readings according to which the dialectical deduction under discussion would employ a demonstrative middle term which in reality does not exist. Philoponus on the other hand, must understand the *toutou* differently, as on his understanding Aristotle is discussing a deduction that does not employ a middle term that is in reality present. Accordingly, I suggest that he takes *toutou* to modify an implicit *sullogismou*. (Cf. Mignucci [1975], 398 who understands Philoponus to take *toutou* to refer to the premise 'no B is A', understood as immediate.)

14. *Topics* 1.1, 100a29-30.

15. Reading *hôste an eiê dialektikos* with R.

16. Philoponus is asserting that whether premises are reputable or scientific, and, accordingly, whether the deduction based on such premises is dialectical or demonstrative, is not a matter of the nature of such premises considered in themselves, but of the epistemic relation that holds between such premises and the one who asserts or hears them in the context of the deduction.

17. Mignucci (1975), 402 asserts that this text is the source of the distinction between natural and unnatural predications found in the later Aristotelian tradition.

18. In a natural predication the grammatical subject refers to the ontological subject.

19. Repunctuating from a question mark to a full stop.

20. Per se accidents (*ta kath' hauta sumbebêkota*) are those attributes of the subjects of the sciences that are not included in their definitions (and hence are accidental) yet necessarily follow from those definitions (for which reason they are *kath' hauta* and are demonstrable). See 1.7 75b1 and *Metaph.* 5.30, 1025a30-4.

21. That is, in addition to substances or other basic subjects of the sciences.

22. *An. Post.* 1.19, 81b30-3.

23. *An. Post.* 1.19, 81b34-82a2.

24. *An. Post.* 1.9, 82a3-9

25. *An. Post.* 1.3, 72b7-14.

26. The Greek phrase *to leukon* ('the pale') can have two senses, 'the paleness' or 'the pale thing'. Philoponus is pointing out that the predication 'the pale is human' is true, and is reflective of the metaphysical order of dependence, if we interpret it as 'the pale thing is human', but that the linguistic predication, in isolation of such understanding, posits as subject 'the pale' in the sense of 'paleness'. For this reason, the predication does not reflect the 'natural' order of predication; it is 'unnatural'.

27. With the Aldine text, reading *proionta dia tôn mesôn*, moving it to after *katêgoroumena*.

28. When Philoponus speaks of a predicate that first proceeds in a certain direction, he imagines one passing through a linear array of all of the middle terms in a full demonstration, and uses the phrase to refer to the middle term that is either the first term one encounters that is another middle term, when one goes in the direction of the minor term, or, when going in the direction of the major term, is the first subject one encounters, of which the next middle term is predicated.

29. 225,4-226,8.

30. Philoponus' commentary makes clear that he takes *ta akra* (the extremes) to be the implied subject of *perainei*. Mignucci (1975), 407 suggests that the implied subject is *ta metaxu* (the intermediate terms).

31. Ross has *perainetai* for the lemma's *perainei*.

32. Philoponus is taking the *touto* ('this') at 82a6 to refer to the situation in which there is an infinite demonstrative chain in any of the three ways indicated in 81b30-82a6: that in which there is a last subject, that in which there is a last predicate, and that in which there is neither last subject nor last predicate. Here he disagrees with Ross (1949), 568, among others, who take it to refer to only the third sort of infinite chain. On the controversy, see Mignucci (1975), 407.

33. There is a lacuna in the text but the sense is clear: by adding a new major term, the previous major term serves as a new middle term, and a new syllogism results.

34. The following sentence shows that Philoponus understands 'this' (*touto*) at 82a6 as referring to the general question whether it is possible for there to be a demonstration that extends to infinity, contra Barnes (1994), 170 who takes it to refer to the more specific question of whether, when the extreme terms of a demonstration are determinate, there can be an infinite number of middle terms.

35. cf. *An. Post.* 1.3, 72b7-15.

36. The first figure syllogistic form by which one would demonstrate a negative premise (Celarent) rests on an affirmative major premise. Hence, every middle term that is inserted serves as an additional minor term, resulting in an affirmative demonstration that this new middle is predicated of the major. Consequently, an infinite negative demonstration would have embedded within it an infinite affirmative one, which Aristotle has shown to be impossible, on account of how there can be no infinite sequences of affirmative predications.

37. Ross reads *eit' ... eit'* .. for *ei t' ... ep'* .. of the lemma; on his reading the whole lemma could be translated 'whether it has infinitely many predicates, and whether both [of the sequences of predicates] are infinite'; on the superiority of Philoponus' reading, see Barnes (1994), 171.

38. Note that here and in the commentary that follows, *katêgoria* must have the sense of 'predicate'.

39. Philoponus is here restricting himself to the context of demonstration. Demonstrable attributes are either per se in the sense of 1.4, 73a34-6, according to which they are included in the definition of their subjects as either genera or differentiae (and hence are not coextensive) or are *kath' hauta sumbebêkota* of the subjects (on which see n. 20). *An. Post.* 2.16-17 is often interpreted to the effect that demonstrated attributes (such as the *kath' hauta sumbebêkota*) convert with the subject, a view which agrees with that of the author of the commentary on *An. Post.* 2 attributed to Philoponus (which likely derives from Ammonius or Philoponus himself) (424,31-426,3; see Goldin [2009], 181 n. 511). Here however Philoponus is not committed to the view that all *kath' hauta sumbebêkota* convert; he rather asserts that all convertible nonsubstantial predicates are *kath' hauta sumbebêkota*. This is because all of the terms in a demonstration are either substantial (the subject, or its genus or differentia) of *kath' hauta sumbebêkota*.

40. Here Philoponus uses his usual term for predicate, *katêgoroumenon*, while clarifying Aristotle's use of the term *katêgoria* as having the same sense.

41. *An. Post.* 1.19, 81b30-3.

42. *An. Post.* 1.19, 81b34-7.

43. *An. Post.* 1.19, 82a3-6.

44. Philoponus is contrasting what he takes to be the order of causation (whereby universal forms are responsible for the characteristics of particular things, with the temporal order of the acquisition of human knowledge, which begins from particulars and ascends to universal truths; see *An. Post.* 2.19, 100a15-b5. Orna Harari points out to us that the assertion that universals are causally responsible for particulars seems in tension with *in Phys.* 12,19-21, where nature is said to put together universals from a collection of particulars. One way of dealing with the issue might be offered by Verrycken's developmentalist account of Philoponus' philosophy, according to which, in at least some passages of both the *An. Post.* and *Physics* commentaries we see the thought of 'Philoponus 2' who comes to reject Plato's Forms (see Verrycken [1990b], 243-54). We reject this account, for the reasons given in the preface. Rather, we suggest, the two passages deal with different kinds of universals. The universal in the sense of a paradigm employed by the Demiurge is first in order of causation; the universal as object of thought is temporally last, following the apprehension of particulars.

45. i.e. A.

46. 'This' (*touto*) must refer to the hypothesis that although the extremes are determinate, the items between them are infinite (225,14-15).

47. Aristotle himself does not here offer the infinite divisibility of a continuum as a model by which one might try to understand how an infinite number of middle terms might be generated, but the disanalogy between demonstrative predications and continua in regard to divisibility is explicitly noted at Aristotle, *Metaph.* 2.2, 994b20-5.

48. The term *anagein*, which here, as elsewhere, we translate as 'classify' literally means 'lead up'. For A to be classified under B is for it to be a term which, alone, or as part of a series, falls under B, and hence has B predicated of it. Here 'classified' might be a misleading translation, as classification, in its primary sense, is a mental process. But Philoponus' point here is that B's status as being under A is already actual, not potential, even if no one mentally takes note of the fact that B has this status.

49. Deleting the first *de* on 226,4 and reinstating the second, which Wallies deletes.

50. Aristotle takes demonstration to primarily show why certain propositions are the case, not to prove that they are true. See *An. Post.* 1.13, 78a2-b13, where Aristotle distinguishes between a scientific proof *that* certain things are the case and a truly explanatory demonstration. See Burnyeat (1981). First principles, for Aristotle, are fundamental truths that are such as do not require explanation; that is not to say that their truth is immediately evident to everyone who considers them. Their truth is rather evident only to those who have mastered the science that they ground. The issue of how we can be *certain* that scientific propositions hold was not foremost in Aristotle's mind; it dominated epistemological debates only in the Hellenistic period. Debates concerning certainty influenced the account of the demonstrative structure of the sciences offered by Ammonius' teacher Proclus. Aristotle had said that the principles of a science need to be taken to be evident to a greater degree than the demonstrative conclusions (one must trust in them to a greater degree [*pisteuomen mallon*, 1.2, 72a33]), but that is not to say that the demonstrative premises themselves, in isolation from the conclusions, provide the ground for one's conviction in them. Rather, Aristotle's point is that the truth of the principles must be evident for the one who employs them as principles. Laymen may not be in such a position. In contrast, Proclus uses the term *autopistos* (self-evident, or self-guaranteeing) to describe the first principles of Euclid's *Elements* (*in Eucl.* 75,14-7 and 179,12-180,3, the latter of which traces the notion of the self-evidence of geometrical principles to Speusippus). The sense here is that such principles are evident by themselves, because of their own nature; they would be so recognised by everyone. The influence of Proclus, through Ammonius, may well be responsible for what R. McKirahan (2008), 4-5; (2009) has shown to be Philoponus' idiosyncratic take on Aristotelian principles, according to which their status as primary consists in how, psychologically, they do not stand in need of justification.

51. As Mignucci (1975), 421 points out, Philoponus' paraphrase, below, shows that he does not seem to be reading *ê ou* at 82a33.

52. Wallies suggests either reading *aph' hou* for *eph' hôn*, yielding the translation, 'So again, in the direction away from those that are infinite, we will never be able to reach [it]' or taking there to be a lacuna which he proposes filling with *ta mesa eis to akron eph' ho estin apeira*, which would yield the translation 'in the case of the middle terms that are infinite, we will never be able to reach the extreme in the direction of which they are infinite'.

53. Philoponus is employing the sense of 'hypothesis' of *An. Post.* 1.10, 76a27-31: a premise which is demonstrable, but is being assumed without being demonstrated.

54. Philoponus is reminding us that so far, Aristotle's arguments for the finitude of affirmative demonstrative changes have been (1) to show that, given a last subject, there is not an infinite sequence of predicates, and that, given a last predicate, there is not an infinite sequence of subjects (81b34-82a2) and that, given any determinate subject and predicate, there is not an infinite number of middle terms (82a3-6); and that this entails the result that (2) on the supposition that there is both highest and lowest predicate, there is no infinite sequence of middle terms (82a20-35); he has not shown that there is in fact a determinate highest subject or a determinate lowest predicate. This is nonetheless assumed, in order to show the consequences.

55. That is, the hypotheses that a given subject term can have an infinite number of predicates, and that a given predicate cannot have an infinite number of subjects, have not been shown to entail impossibilities.

56. The syllogistic form is Celarent.

57. Ross has *hôste epei hê epi to anô istatai hodos, kai hê epi to A stêsetai* for the lemma's *hôst' epei ê epi to katô istatai hodos, kai ê epi to anô stêsetai*, which could be translated, 'So since the path upwards stops, the path to A, too, will stop.'

58. The 'path downwards' is, as Philoponus points out, one of a process of supplying additional middle terms in the direction of the minor term. This involves the generation of a series of first figure affirmative syllogisms by which the minor premise of Celarent is progressively mediated. But, as Ross points out, 571-2, what Aristotle needs to show is that there cannot be an infinite sequence of first figure negative syllogisms of the form Celarent mediating the major premise. This is why Ross amends the text, on which see the previous note.

59. The syllogistic form is Camestres.

60. This would be through the syllogistic form Cesare.

61. Philoponus is pointing out that a universal negative predication can be proven in the second figure either through Camestres or Celarent. In the case of Camestres, the minor term is itself a universal negative predication (and can thus itself be proven via a new middle term, either through Camestres or Cesare) and in the case of Cesare, the major premise is a universal negative predication (and again, given a new middle term, can itself be proven either through Camestres or Cesare). Philoponus tells us that a syllogism of the form Cesare can be 'increased', that is, made longer through the insertion of middle terms, through subordinate syllogisms of either the form Camestres or Celarent. The present discussion, however, concerns whether a negative universal demonstration can possibly rest on an infinite sequence of affirmative universal predications. Hence he does not pursue the case in which a negative universal demonstration is indefinitely lengthened by subordinate syllogisms of the forms of both Camestres and Cesare, with the one form always, at some point, followed by the other, as in such a case one could not arrive at an infinitely long sequence of affirmative predications. We are indebted to Orna Harari for help on this passage.

62. In other words, whether one indefinitely augments a syllogism of the form of Camestres or of the form Cesare one will generate an indefinite sequence of affirmative universal predications through the insertion of new middle terms within the initial universal affirmative predication, by virtue of an infinite sequence of syllogisms of the form Barbara, and this will entail an infinite sequence of subordinate universal affirmative predications.

63. By 'the initial middle term' Philoponus means the first middle term that is taken in increasing the universal affirmative premise with which one begins. For in the case of Cesare, it is the minor premise that is universal and affirmative, and hence the second middle term that is taken will be less universal than the first.

64. This would be the case in which it is the negative universal premise that is 'increased' through the insertion of subordinate syllogisms of the form Camestres or Cesare.

65. Above (82b13) Aristotle offered as an example of a second figure negative universal syllogism an instance of Camestres: B belongs to no C, B belongs to all A; A belongs to no C. Here he discusses how the negative, minor premise of that syllogism (B belongs to no C) can be proven through the insertion of another middle term, D, and another instance of Camestres: D belongs to no C, D belongs to all A; B belongs to no C. Here the second figure is called the 'middle figure'.

66. See 82a39-b3. Aristotle there explicitly clarifies the technical sense he is here giving 'first and last'; the senses of 'upwards' and 'downwards' must be gathered from the context.

67. It is not at all clear why Aristotle moves on to discuss the possibility of infinite sequences of predication in third figure syllogisms, since, as Philoponus points out, all of their conclusions are particular, for which reason none of them shall serve as demonstrations. Barnes, 173, accordingly excises 82b21-8. Philoponus tells us that Aristotle discusses this *ek periousias*. Orna Harari has suggested to us that the phrase here means 'for the sake of completion'. On her proposal, which we accept, Philoponus sees Aristotle as having already gone most of the way in discussing the full range of possibilities by which there might be infinite sequences in syllogisms, and he takes Aristotle to be figuring that he might as well finish up the account.

68. Philoponus understands 'C does not belong' to be short for 'C does not belong to any B' (*oudeni tôi B*) and consequently takes the third figure syllogism in question to be Felapton, while Ross (1949), 571, among others, understands it as short for 'C does not belong to every B' and accordingly understands the syllogism in question to be Bocardo. See Mignucci (1975), 431-2.

69. Philoponus slips; he offers an example of an invalid third figure syllogism. As he himself pointed out at 232,3-4, no third figure syllogisms have universal conclusions.

70. Ross has *peperasmenakis* for the lemma's *peperasmenôs*.

71. Accepting Wallies' reading *peperasmenakis* for *pollakis* of the manuscripts, on the basis of the lemma.

72. Philoponus takes *hodos* to refer to 'syllogistic figure'; Mignucci (1975), 436 argues against this, taking it to refer to a valid syllogistic form.

73. Philoponus' commentary groups 82b35-6 with what follows, hence he would regard the chapter division as at 82b35, not at 82b37, as in Ross's text.

74. These are the most specific species, that are indivisible in the sense that there are no subspecies that fall under them (although they can, in a sense, be divided into the particulars that make them up). They are predicated of these particulars, but not in the context of demonstrations.

75. Alexander of Aphrodisias, Peripatetic philosopher and one of the earliest commentators on Aristotle, active in the late second and early third century AD.

76. An alternative translation is 'which he primarily employed'.

77. In what follows, Philoponus preserves for us a dispute concerning the sense in which at 1.21, 82b35 and 1.21, 84a7 Aristotle calls the arguments offered at 1.22, 82b37-84a6 'formal' (*logikos*). According to Philoponus' report, Alexander (followed by Ross [1949], 573, 575-9) understands the term here as having the sense of 'dialectical'. (See *in Top.* 30,12-14, where Alexander points out that Aristotle often employs the term with this sense.) On this understanding, the arguments here are dialectical insofar as they are taken to rest on an assumption – the existence of indemonstrable definitions – that is granted insofar as they are generally accepted, even though they have not (in this context) been proven. (Against Alexander, one might ask why, on this understanding, Aristotle calls the argument at 82b37-83b31 'more formal' than that of 83b32-84a6? Is it that that thesis that there are definitions is somehow more foundational than the thesis that there are demonstrations?) Ammonius, on the other hand, took the existence of definitions as self-evident, that is, as itself a first principle, not requiring proof. He rather takes Aristotle to be calling his argument *logikos* insofar as it rests on general logical considerations concerning predicative chains – its force derives from its general logical structure, not the fact that Aristotle's initial argument, as he initially presents it, concerns essential, substantial predications (a point that Aristotle explicitly acknowledges when he extends his line of argument to definitional chains concerning items outside of the category of substance). (On this sense of *logikos* see e.g. *GA* 2.8, 747b27-30.) Philoponus adds a second reason for calling the argument *logikos* – that definitions are wont to be employed in arguments that are *logikoi* (in the first sense, presumably), and the argument at hand rests on a consideration of definitions.

78. i.e. Ammonius; see n. 5.

79. Ammonius is alluding to Hellenistic epistemological debates. In order to defend their view that the sage is not subject to moral error, the Stoics posited a certain variety of 'appearance' (*phantasia*) that not only presented things as they truly are, but was self-certifying, with a certain characteristic that, for those able to recognise it, guaranteed that it presented things as they really are, and thereby gives the sage the green light to assent to, and act in accordance with it. Apprehension (*katalêpsis*) is such an assent, to this sort of impression, called 'apprehensive'. (It is a *katalêpktikê phantasia*.) Academic and Pyrrhonian sceptics both denied that there are such impressions, and, accordingly, that there is such an assent. Ammonius is considering one's awareness of a definitional first principle as apprehensive. In so doing, he is interpreting Aristotle along the lines seen earlier in the present commentary, according to which first principles are not so much foundations for understanding why a demonstrative conclusion is the case (the usual contemporary understanding of Aristotle's notion of a first principle), but self-evident truths grounding justification that demonstrative conclusions hold. See n. 50.

80. Hence, Ammonius argues, Aristotle's first argument does indeed posit an indemonstrable principle as a basic premise, for which reason it is not, pace Alexander, to be considered dialectical.

81. Ammonius seems to infer that the self-evidence of definitions entails that it is self-evident that there are definitions.

82. This remark is perplexing because Alexander did not say that Aristotle's argument was dialectical because it rests on premises that were persuasive but not true, but because its premises, though true, indeed, demonstrable, were put forward on the basis of their persuasive character.

83. On the interpretation of Ammonius, Aristotle calls his arguments here *logikos* because they apply not to a specific kind (in this case, demonstrative chains) but to more than one kind (here, to all predicative chains). Accordingly, they fall short of the standards of demonstration, which falls under a science that concerns only one kind. Instead they appeal to general formal or logical considerations.

84. 82b37-83b31.

85. Philoponus is alluding to how the term *logikos* ('formal') can have the sense of 'logical' (see for example 1.6 of the present commentary).

86. The term translated as 'category' is *katêgoria*, translated as 'predication' in the previous sentence. The shift in sense is not as abrupt as might first appear; see *Topics* 1.9, where Aristotle initially offers a classification of ten categories as kinds of predication (*ta genê tôn katêgoriôn*; 103,21-2); from this, he proceeds to offer an associated list of ten kinds of predicate (103,29).

87. This is similar to the account of why this argument is called *logikos* offered by Ammonius (234,4-13). But while Ammonius said that the argument is 'formal' as it in principle can apply to more varieties of predications than per se ones, Philoponus says that it is formal because it applies to more than one variety of per se predication.

88. 1.19, 81b24-9.

89. Here, the distinction between substance and accident is offered as an ontological distinction, whereby an accident is any being that is not a subject, but inheres in a subject. Aristotle, in contrast, is distinguishing not between kinds of beings, but kinds of predications: an accidental predication is one where the true subject of the predication is either unmentioned (as it is referred to by means of one of its predicates) or serves as predicate. The two distinctions are not the same. For example, quantities are not substances, but, in the context of mathematics, they serve as subjects. 'The line is straight' is, in the context of geometry, a 'natural predication'. Philoponus recognises this, in passing, at 235,29-236,8.

90. A grammatical subject, of course, does not have a nature in the strict sense, since it is not the sort of thing that can move. But it does have a role that it is supposed to play, which, metaphorically, can be said to be in accordance with its nature. It is this that Philoponus appeals to in distinguishing between those instances in which a term is used in accordance with this role and those in which this is not so. What Philoponus calls a natural predication would be a predication in which the grammatical function of the subject is in accordance with the essence of its referent.

91. 'Happens' renders *sumbebêke*; the passive participle of which is *sumbebêkos*, the term rendered 'accident'. An accident is what 'happens' to be the case concerning something; it is included in, or necessitated by, its essence.

92. In such a case, the predicate is a definitional predicate of the subject, so that, even though the subject does not refer to an ontological basic subject, the subject is still the proper subject of the predicate, as the predicate exists only insofar as it is a feature of the subject (which itself exists only insofar as it inheres in an ontological subject). In the terms of the *Categories* (3, 1b10-15), the predicate is 'said of' the subject.

93. Philoponus moves from considering an accident as a kind of being, to considering it as a kind of predicate (understood as an accident only contextually, in relation to its subject). 'Colour' is an accident of 'human being' (as it is a nonsubstantial characteristic that is not part of the essence of the substance that is its ontological subject) but it is part of the essence of 'white', as white is defined as a colour dispersive of light. So, Philoponus suggests, in the context of considering the predication 'white is a colour' alone, 'colour' is not to be considered an accident.

94. Philoponus' distinction between two ways in which an accident is predicated of an accident will not stand. For in the case of a predication such as 'the white is hot' (indicating that something, such as a human being, that is white is also hot) a quality is predicated of a quality, but 'hot' is not part of the essence of 'white', for which reason 'white' does not refer to a true subject for 'hot', and the predication is to be considered unnatural.

95. Philoponus understands *hoper* ('precisely') as implicit before the first occurrence of *leukon* ('pale') at 83a7.

96. Aristotle often uses the term 'precisely' (*hoper*) with the sense 'essentially': to say that A is precisely B is to say that B is the essence, or a part of the essence, of A. See LSJ, s.v. b. The phrase *outh' hoper leukon ti* at 83a7-8 is often so understood, as having the sense 'nor insofar as it is the essence of a particular shade of white'. (Here *ti* indicates not an ontologically particular predicate, in contrast to a universal predicate, but a specific *kind* of predicate). See Ross (1949), 581; Urbanas (1990), 101. But Philoponus cannot be understanding the phrase in this way, as at 237,10-11 below he says that we say that the subject is *hoper* the predicate when a subject and predicate are (merely) coextensive, as when the predicate is a property of the subject. But what is Philoponus' understanding of the phrase at 83a7-8? This is not easy to determine. We tentatively offer the following. Philoponus seems to be employing *hoper* in its looser, nontechnical sense ('which very thing'), according to which A is B if B precisely identifies A in some relevant way. (Cf. LSJ s.v. A and Urbanas [1990], 101-8.) Philoponus gives three examples of the *hoper* locution, as he takes it to be used in the lemma. The first is 'human being is precisely receptive of intellect and science'. The idea here cannot be that 'receptive of intellect and science' is the essence of 'human being' (for that is 'rational animal, while 'receptive of intellect and science' is a mere property; see 223,11-12) but that in some unspecified context, 'receptive of intellect and science' serves to precisely identify what is human. Philoponus' offers three other examples. One is 'the one approaching is a human being'; in which case the bearer of an accident is precisely identified as a kind of substance. The other two are 'this particular one (*tode*) is precisely white' and 'this particular one is precisely a particular white'. Again, Philoponus is alluding to, but not specifying, a context in which 'white' (or 'a particular shade of white') would serve to uniquely identify the referent of 'this particular one'. Philoponus is interpreting Aristotle as saying that it would be false to say that 'what is a stick is precisely white' along these lines. In the relevant context (which would here be that of the demonstrative sciences), 'white' does not uniquely and appropriately serve to identify the stick. Why not? Philoponus goes on to remark that, insofar as the stick is the persisting substrate for the white colour (and not the other way around), the stick must here be referred to *as* a subject. An unnatural predication, however, identifies the attribute as subject.

97. As the next sentence makes clear, it is the species 'human being' that is the subject.

98. Accepting Wallies' omission of *gar*.

99. On this basis, Barnes (1994), 176, suggests that Philoponus read *hoper leukon on outh' hoper leukon ti* for *leukon on outh' hoper leukon ti* at 83a7.

100. We translate in accordance with Philoponus' interpretation of the text, below. Ross, 581, understands *hoper kai egeneto* at 83a17 as 'which is what we made it in our assertion'.

101. Again, Philoponus is endorsing a usage according to which we can say that 'the pale is precisely cultured' (which shows that he is not here understanding *hoper* as 'essentially'); he is rather saying that the context in which we would precisely identify the pale as cultured is not a context employed in the sciences.

102. In both cases, Aristotle invents a term ('accidental predication', 'relation') where there was none before.

103. The *aei* does not apply to the verbal phrase 'is predicated'; it rather applies to the whole statement that nonaccidental predications are predications in the strict sense.

104. *DC* 1.2.

105. *DC* 1.5-7

106. *DC* 1.8-9.

107. *DC* 1.10-12.

108. Aristotle's argument for the finitude of demonstrative chains rests on the assumption that all of the predications that feature in demonstrations are 'natural,' according to which either a nonsubstance is predicated of a substance, or a feature that figures in the definition of a nonsubstance is predicated of it. But here Philoponus wonders about the demonstrative conclusions of geometry, in which the subject is a quantity (which is not a substance).

109. Philoponus' response to the objection is inadequate, as it ignores the distinction that Aristotle clearly makes between the definitional attributes of geometrical entities and their demonstrated attributes, the *kath' hauto sumbebêkota*. It is not at all evident how no definitional demonstrated attributes are to be understood as identified with definitional attributes of the basic entities studied by geometry. For a suggestion concerning how this might work, see Goldin (1996).

110. It was not obvious in the case of the mathematical sciences that sciences study how an accident is predicated of an accident, but, in contrast, it is obvious that no science studies unnatural predications, of the kind seen in the examples Philoponus is about to give. What is the distinction between the mathematical case of an accident apparently predicated of an accident (a quantity of a quantity) and the example of the unnatural predication (three cubits tall is pale) that Philoponus gives here? Perhaps the present example differs insofar as it is a case in which an item of one category is predicated of an item in another (nonsubstantial) category.

111. Philoponus does not make clear exactly what the puzzle is. Is the puzzle that demonstrations will have major terms that are not definitional of the subject? Goldin (2010) argues that the lost *Posterior Analytics* commentary of Alexander of Aphrodisias recognises that the demonstrative scheme that Aristotle is setting up involves a major problem of how nondefinitional predications are to be syllogistically deduced from definitional ones, and offers a solution, in which he was either not comprehended or not followed by Ammonius and Philoponus. Perhaps the current passage is an instance of a phenomenon seen fairly often in the tradition of ancient Aristotelian commentary: a puzzle or topic of discussion is dutifully handed down from earlier commentary to later one, long after its point or relevance has been forgotten. On this see Lloyd (1990), 1-35.

112. Again, neither the problem with which Philoponus is dealing nor the solution is clear. Following up on the suggestion of the previous note, we suggest the following. The problem concerns how a natural scientist (*phusikos*) dealing with a moving body like the earth, will be concerned with a mathematical quantity like sphericity. The definition of a natural body will deal with its natural characteristics (the characteristics by virtue of which it will have an internal principle of change or rest); how will mathematical simples, from which a quantity can be deduced, be included in the definition of such a thing? Philoponus' solution is the following: the mathematical major term is predicated definitionally (that is, 'primarily' or 'per se'; see 1.4 73a34-6) of a mathematical kind (such as 'shape of the earth'); this other kind ('shape of the earth') is then shown to inhere in the minor term, on the basis of the science (here, physics or natural science) that concerns the proper genus of the minor term. The mathematical demonstration and the physical demonstration are bridged by the fact that the subject, earth, can be considered from both a physical and a mathematical point of view. This reconstruction of Philoponus' point of view is in line with the Alexandrian account of the demonstration of the *kath' hauta sumbebêkota* offered in Goldin (2010).

113. It is a fact of natural science that the earth, as immovable centre of the cosmos, is unique. The shape of the earth, on the contrary, is a subject of mathematics (as 'shape' belongs to the category of quantity); as such, there is no mathematical demonstration that there is only one entity with such a shape.

114. This, we suggest, was the view found in the commentary of Alexander; see n. 113 above. Note that Philoponus himself is doubtful concerning the suggestion, and offers another, which, in effect, leaves Aristotle without a way of demonstrating the *kath' hauta sumebêkota* in the context of his syllogistic.

115. Philoponus here as elsewhere understands *hoper ... ti* as 'precisely the particular'. Ross in contrast here takes *ti* to be in interrogative: 'just what that subject is a species of'.

116. This is an example of identifying the subject as 'precisely the predicate'.

117. This is an example of identifying the subject as 'precisely a particular predicate'.

118. In the case in which we predicate an essential attribute of a substantial term, it is not the case that the subject precisely and uniquely identifies the predicate; because the subject may well be less coextensive than the predicate, there are aspects and instances of the predicate that the subject term will not serve to identify. But the subject is precisely the predicate, as the predicate term will precisely identify the subject, albeit at a higher level of generality.

119. Philoponus takes *ousian* at 83a24 to have the sense of 'substance' in contrast to 'essence', which is the sense it is usually given; see Ross (1949), 574.

120. On Philoponus' construal, Aristotle is saying that in the case of a predication 'S is P' when S is a substantial term and P is an essential attribute of S, the term P refers to the very substance of which P is predicated (albeit, perhaps, at a different level of generality than does the term S).

121. Reading *to katêgoroumenon* with a.

122. That is to say, the predicate, as described, either generically or as belonging to an *infima species*, does not precisely and appropriately identify the subject. In the first case, the subject is precisely identified as belonging to a kind; in the latter, the level of specificity of the kind is relevant to the precision by which the subject is identified as belonging to that kind.

123. Within the *Categories* Aristotle referred to nonessential predicates as 'being in' their subject of which they are predicated, but, as Philoponus indicates, he does not maintain this terminology and here, as elsewhere, spoke of them as 'being predicated of' their subject.

124. Aristotle himself uses the phrase *kath' hauto* (*per se*) to refer to a subject that independently exists at *An. Post.* 1.4, 73b5-8.

125. Philoponus here offers a 'since' clause, follows it with a long parenthetical remark, and then loses his train of thought, as he reprises another 'since' clause following the parentheses.

126. Neoplatonists referred to the Ideas, in their role as generative of their images in the sensible realm, as *logoi*. See Lloyd (1990), 92-3.

127. This is a *hapax legomenon*, a unique occurrence of a term in the Greek corpus. Etymologically, it has the sense of 'handling [an instrument] in advance, in all kinds of ways'. Its sense is either that of tuning strings, or running difficult lines in advance of a performance, as one hears members of an orchestra do, prior to the arrival of the concertmaster. The metaphorical sense is clear enough: a display of virtuosity without sense or any real effect.

128. At Plato, *Theaet.* 165A, Theodorus likewise considers Socrates' arguments as *psiloi logoi*, although there the sense is 'bare arguments', that is, arguments that are abstract, not concerning specific things.

129. That is, they exist only as inherent within matter.

130. The distinction is between some higher level matter, which has enough internal organisation to constitute a body, and absolute prime matter, which like the Receptacle of the *Timaeus*, serves as absolute substrate, but does not have enough form to count as 'a body.' (Philoponus' thought on the distinction underwent serious revision, on which see de Haas [1997].) Philoponus' interpretation of Aristotle's remarks is problematic: the inherence of forms in matter should be irrelevant to Aristotle's argument, first, because the notion of matter is absent from the *Posterior Analytics*, and second, the minor terms of demonstrations will not be terms designating instances or kinds of matter (except in the case in which one is studying kinds of matter, in which case it is the form of that variety of matter that is the subject of demonstration).

131. An alternative translation is 'the demiurgic accounts are ideas of things'.

132. The reference is to Aristotle, *Metaph.* 12.10, 1075a13-15, 1076a4, understood in a peculiarly Neoplatonic manner, identifying Aristotle's Unmoved Mover with Plato's Demiurge.

133. As Wallies indicates, there must be a lacuna in the text. The sense is clear enough: if a doctor is ill and then is treated, this will not occur spontaneously, but by virtue of a regimen that is intelligible, and accordingly subsists not in the doctor's body (as do the motive principles of a human being) but within the doctor's mind. It is for this reason that the doctor's healing himself is not a natural process (cf. *Phys,* 2.1. 192b23-7) but requires the artifice of the art of medicine, just as, according to Plato's *Timaeus*, the organisation of the cosmos requires the artifice of the Demiurge, in accordance with the intelligible Forms.

134. Aristotle nowhere identifies the Unmoved Mover as a Demiurge that is responsible for the order and organisation in the cosmos, He does say that the Unmoved Mover is an intellect that thinks himself (*Metaph.* 12.9, 1074b21-35), but nowhere explicitly says that it thereby knows all things.

135. Aristotle, *DA* 3.4, 429a27-8.

136. In the early *in DA* 37,16-38,16, Philoponus himself argued against interpreting Aristotle as disagreeing with Plato concerning the Forms. See pp. 2-3 above. Some of the same texts were there adduced as evidence as are here mentioned by the reconcilers; it is evident that they include Philoponus' younger self, as well as Ammonius. In his *DA* commentary, Philoponus himself did not cite *DA* 3.4, 429a28-9 ('Those who say that the soul is the place of Forms speak rightly') in his defence of the harmonisation of Plato and Aristotle on the Forms, but Asclepius, *in Metaph.* 69,17-22 and 167,14-34, provides evidence that this passage was cited to this effect by Ammonius. Both of these passages from Asclepius also adduce *DA* 3.4, 430a14-15, 'there is an [intellect] that is what it is by making all things', taken to be an allusion to the creative function of the divine *logoi*, as evidence for Aristotle's belief in the divine *logoi*. This too presumably derives from Ammonius. See Verrycken (1990a), 220.

137. That is, as demiurgic ideas.

138. That is, others were mistaken in their interpretation of Plato's account, so they were led to attribute to Plato Forms of the sort that would be subject to Aristotle's criticisms.

139. Philoponus is thinking especially, but not only, of *Metaph.* 1.9.

140. Following Wallies' conjecture of *hôs paradeigmata eikonôn* for *eis paradeigma ex ekeinôn*.

141. For the distinction between 'objection' and 'counter-objection' see Philiponus, *in Cat.* 81,10-13. An objection to a puzzle (*aporia*) is an argument indicating that the line of argument behind the puzzle is not to be accepted, and that accordingly the puzzle has no force. A counter-objection accepts the line of reasoning behind a puzzle, but shows that the puzzle has no force in regard to the matter in regard to which the puzzle was directed.

142. This is an objection: it accepts the *aporia* to the effect that some predications are unnatural but shows that this does not call into question the thesis that there are no predicative chains of infinite length.

143. This is the counter-objection: it refuses to accept the initial argument that some predications are unnatural.

144. In contrast, Ross, 581, takes *poiotês* at 83a36 to signify an attribute in any category.

145. For the definition of white as 'dispersive of vision', see Plato, *Tim.* 67E and Aristotle, *Top.* 3.5, 119a30-1.

146. 234,13-235,1.

147. Philoponus says that he is proving that unnatural predicative chains cannot be infinitely long, but he has proven much less than that. He starts by considering an unnatural predication such as 'the white is stick' as a syllogistic premise. He first dismisses the possibility that white is an element in the definition of stick. (This would be the case in which the subject is itself a *kath' hauto* predicate of number in the second sense outlined in *An. Post.* 1.4, 73a37-b3. An example of such a premise might be 'Odd is number'.) An infinite sequence of predications cannot be generated in this way, for that would entail an infinitely long definition, which has been shown to be impossible. He then considers the possibility that white is not present in the definition of stick, that is, that it is an accident. What Philoponus would need to show (but is not able to) is that there is no accident that is predicated of white such that there is in turn an accident predicated of that, and so forth, to infinity. Instead, he converts the initial accidental unnatural predication to an accidental natural one ('the stick is white') and then shows how we cannot generate an infinite sequence of definitional predications of the predicate.

148. The passage is most naturally translated as 'the same thing cannot be a quality of itself' but this – that for no quality P is P a quality of P – is not what is at issue.

149. It is not possible for something (e.g. human being) to be the genus of the same thing (e.g. animal) that is its genus.

150. It is not possible for something (e.g. stick) to be a quality of the same thing (e.g. white) that is its quality. Philoponus switches from saying that there cannot be a quality of a quality in general to saying that a subject cannot be a quality of its quality and a species cannot be a genus of its genus. At 245,7 he returns to the issue of whether there can be a quality in general of a quality in general.

151. It is not clear what the reference is, but the contention is not controversial. At *Cat.* 3 1b10-15, Aristotle indicates that in the case of a definitional predication (in which the predicate is 'said of' the subject) everything that is 'said of' the predicate is 'said of' the subject. So if substance is definitionally predicated of the predicate, and the predicate is predicated of a quality, the quality will be a substance, which is absurd.

152. Philoponus, following Aristotle, has already denied that an unnatural predication is a predication in the strict sense. Here he adds to that, the predication of a genus of a species. Aristotle said that in a proposition, one item must be predicated of another (*ti kata tinos*) (*Int.* 5 17a20-2). Philoponus draws the conclusion that, in the strict sense, an item cannot be predicated of itself at all.

153. If an essential predication is not a predication in the strict sense, then, it would seem, it could not serve as a demonstrative premise. But definitional predications are among the immediate premises. So the fact that a predication is not a predication in the strict sense should not tell against it playing a role in demonstration. This should call into question the validity of Aristotle's argument, which banishes unnatural predications from playing a role in a demonstrative predicative chain. Philoponus does not seem to be aware of the difficulty.

154. Accepting Wallies' deletion of *kata sumbebêkos*.

155. Philoponus says that this other disjunct must be understood as following the first sentence of the lemma.

156. Aristotle is here summarising what he has proved. He has proved that accidentally predicative chains cannot be infinitely long. Hence he is to be understood as indicating that in his summary.

157. *An. Post.* 1.22, 83b10.

158. Philoponus takes the ten categories to be highest genera. The sort of descriptive account that Philoponus has in mind is presumably the sort that we see in Aristotle, *Cat.* 1-5.

159. The term *noein* can refer either to conceiving or (on the Neoplatonic understanding of Aristotle) to having an immediate nondiscursive knowledge of principles. Philoponus is indicating that Aristotle here has the second use in mind. The point is not that one cannot conceive of an infinite (for one can, insofar as the term *apeiron* is meaningful) but that within the human mind there cannot be an infinite number of actualised grasps of essences. See 249,15-18.

160. The objection is against the possibility of unnaturally predicating a genus of a thing.

161. That is, the other categories besides substance.

162. *An. Post.* 1.20.

163. *An. Post.* 1.21.

164. *An. Post.* 1.22, 83b1-8.

165. *An. Post.* 1.22, 83b10-17.

166. 235,2-7.

167. Ross reads *ti tôn toioutôn*, instead of *tôn toioutôn*; the text as Philoponus presents it requires that *ti* be understood.

168. Ross has *dê* for *de*.

169. Reading *tou leukou to khrôma* with R.

170. Aristotle is distinguishing between two kinds of attributes that are accidental (included in the essence): those that are per se accidents, and those that are not. The per se accidents (*kath' hauta sumbebêkota*) are predicates that are demonstrable of the subject, but not predicated of the subject in its definition. Philoponus here, on the other hand, understands a per se accident to be a nonsubstantial predicate that is in the definition of another nonsubstantial predicate (and hence is per se) but is not in the definition of the subject in which it inheres. Philoponus' interpretation does not make clear why Aristotle thinks that pointing out that some nonsubstances are predicated per se is relevant to the task at hand (showing that all nonsubstances are predicated of substances) while on the standard account (like that of Barnes [1994], 179) the remark has a point: nonsubstantial predicates will indeed play a role in demonstration, but they will either be regarded as ultimate subjects (as are mathematical simples) or will be regarded as predicated of such subjects.

171. In other words, every nonsubstantial predicate has its own essence, yet the subject for that predicate is something different, with a different essence.

172. As Philoponus notes below (225, 268), there are variant manuscript readings here. The lemma has *all' atta kath' heterou* at 83b24, but as Philoponus reports, other manuscripts have *touto kath' heterou*. Ross accepts the alternative text that Philoponus reports.

173. Wallies suggests filling in the lacuna with *epei to hen ê kath' henos*, according to which the opening of this sentence is to be rendered 'Given that one thing is predicated of one thing or of many ...'. Philoponus' previous arguments against the infinitude of demonstrative chains rested on the assumption that within demonstrations, predicative chains are linear; a single demonstration does not express branching lines of predication. For example, even though it might be the case that A is B, and B is C, and also that A is D and D is E, there will be one demonstration to the effect that A is C and another that A is E. In what follows, then, Aristotle is taken to be aiming to prove that this assumption holds.

174. We take this to refer to the phrase *hen kath' henos ... huparkhein lekhthêsetai*.

175. It is not clear why the fact that many things are not predicated of one is said to entail the fact that one thing is not predicated of many.

176. Philoponus supplies the crucial qualification 'to infinity', which, as Barnes, 179, points out, is missing in Aristotle's text.

177. Even if one does not grant that demonstrations must be restricted to nonbranching lines of predication, Aristotle's conclusion that demonstrations cannot embrace an infinitude of predications still follows. For the multiplicity of predications are themselves comprised of a finite number of fundamental units. Hence, since the product of two finite numbers is finite (cf. 221,6-9), even a branching demonstration involves only a finite number of predications.

178. 253,15.

179. The three hypotheses were the three possible cases that were posited by Aristotle, to see whether there follows a finite predicative chain. The first, presented at 1.19, 81b30-3 and discussed at 1.22, 82b36-83b12, was that there is a determinate ultimate subject (220,16-20). The second, presented at 1.19, 81b34-82a2, was that there is a determinate ultimate predicate (220,22-4). The third, presented at 1.19, 82a2-6 and discussed in 1.22, 83b13-24, was that there are determinate ultimate terms on both sides (220,26-221,4).

180. Ross has *esti* for *estai*.

181. See 233,4-234,31.

182. 233,32-234,4.

183. The reference is likely to *An. Post.* 1.3.

184. Against Alexander, Philoponus takes *logikôs* at 82b35 as having the sense of 'holding generally, in regard to any subject matter'. Accordingly, he denies that Aristotle is using the term in regard to the argument at 83b32-84a25.

185. This was in *An. Post.* 1.13.

186. See *Metaph.* 12.9, 1072b14-30 for Aristotle's discussion of *noêsis* as the best activity. See also *An. Post.* 2.19, 99b26-34 on the superiority of *noêsis* over demonstrative science in respect to precision.

187. The Stoics took perceptions to give rise to certain beliefs shared by all rational beings, which they call 'common notions'. Following Proclus (on whom, see Martijn [2010], 112-14) Philoponus here, as elsewhere, uses this phrase to refer to Aristotle's first principles by reference to their epistemological status; cf. McKirahan (2008), 115-16 nn. 17, 18; (2010).

188. In *DA* 3.4-5 Aristotle offers an account of how intelligibles are grasped by *nous* that is parallel to his account of how simple sensibles are grasped by sensation.

189. See 1.2 72a32-6.

190. On apprehension, see n. 79.

191. As Mignucci (1975), 484 points out, Philoponus sees that Aristotle has not here offered a proof strong enough to show that there is no infinitely long demonstrative chain, for which reason he takes the present proof to be limited in its scope.

192. Omitting *tou eidenai* with Wallies.

193. Philoponus takes *ek tinôn* to be synonymous with *ek hupotheseôs.*

194. Philoponus is relating the use of *hupothesis* here to that of 1.10, 76b27-34.

195. This is an odd example for an Aristotelian commentary. Presumably the proof would show how an afterlife follows from the world's being run by divine providence, for providence requires that the righteous and the wicked would receive their just deserts, and it is evident that this does not occur while they are alive.

196. Philoponus has *diexelthein* for Ross's *dielthein.*

197. 1.4, 73a34-6.

198. 1.4, 73a36-b3. Odd is per se, in this sense, of number, since 'number' is in the definition of odd; snub is per se of nose, since nose is in the definition of snub (see *Metaph.* 7.4, 1030b14-27). It is controversial what role Aristotle takes such predications to play within demonstrations; Philoponus himself does not discuss this crucial issue in his commentary.

199. *An. Pr.* 1.1, 24a10-11. On debates in late antiquity on what it means to say that the study of the analysis of syllogisms is for the sake of demonstration, see Goldin (2010).

200. Accepting Wallies' insertion *huparkhei to prôton, en tôi.*

201. If P is per se, in the second sense, of S, S is in the definition of P. If Q is per se, in the second sense, of P (and is therefore per se, in the second sense of S, too, since, because it has P as a subject and P has S as a subject, Q has S as a subject too, and would have S in its fully explicit definition), then P is in the definition of Q. Since S is in the definition of P, both S and P are in the fully explicit definition of Q. Aristotle's argument is that this process cannot go on indefinitely. Philoponus does not make clear how this result, that the definitions of predicates that are per se in the second sense are finite in number, bears on the question of why there cannot be an infinite number of predications within a demonstration, for it is not at all clear how such per se predications play a role in demonstrations.

202. The objection is as follows. Aristotle is not here, as before, envisaging an infinite process of the insertion of middle terms. Rather, he imagines adding on ever more major terms. The generation of terms that is per se in the second sense would be potentially infinite, and Aristotle has no objection to a potential infinity. Thus, he begins with 'number is odd', which is per se in the second sense (as odd is defined as a kind of number) and then he adds 'odd is prime' (as prime is defined as a certain kind of odd number) and so on, to infinity. The objection runs that, whichever of these new major terms are considered, there will be a finite number of terms within its definition, and no absurdity results.

203. Philoponus argues that an infinite regress of predicates that are per se in the second sense does indeed result in an infinite number of terms predicated of the minor term. But it remains unclear why this is not a mere potential infinity of terms. In the case of predications that are per se in the second sense, the subject is taken in the definition of the predicate; the predicate is not taken in the definition of the subject. So it is not as though the minor term is thereby rendered unknowable – although number is in the definition of each of them, it is not the case that each of them is in the definition of number.

204. That is, the absurdity that the objector sees the argument purportedly leading to, that each predicate that is per se in the second sense has an infinite number of definitional predicates.

205. Ammonius argues that an infinite regress of predicates that are per se in the second sense would result in infinitely long definitions of the minor term, the ultimate subject of the per se predicates.

206. Philoponus realises that Ammonius' account of the reductio is unsatisfactory, as the infinite regress of predicates that are per se in the second sense does not, in fact, lead to any infinitely long definitions.

207. Philoponus' point is that the existence of an infinite number of attributes that are demonstrated to hold of a subject (those which are *kath' hauta sumbebêkota* in respect of it) is incompatible with the thesis that complete scientific understanding of a kind is attainable, since these attributes could not all be scientifically known to hold of the subject. He apparently is identifying the *kath' hauta sumbebêkota* with per se attributes of the two varieties Aristotle distinguishes at 73a34-b3 and in the passage that follows the lemma. But Philoponus did not make any such identification in his discussion of 73a34-b3; see McKirahan (2008), 133 n. 332. A problem with Philoponus' interpretation is that Aristotle nowhere says that scientific understanding regarding a subject rests on knowing everything demonstrable of that subject. Mathematics apparently presents many counterexamples to such a view, as, for example, the features of regular polygons can be understood as demonstrable of their sides (line segments), but there are infinitely many regular polygons. Mignucci (1975), 490 suggests that Philoponus here has in mind *DA* 1.1, 402b16-25, where Aristotle says that knowledge of the *kath' hauta sumbebêkota* of a subject contributes to knowledge of its essence, but then (rightly) rejects it as providing adequate support for Philoponus' interpretation.

208. Reading *endekhesthai* for Wallies' *endekhethai*.

209. Mignucci (1975), 491 argues against the interpretation of Alexander reported here by Philoponus, by pointing to 84a24, where, on his interpretation, Aristotle asserts that the terms that are *kath' hauta* in the second sense are all convertible with each other and with the subject. But at 262,2-8 Philoponus interprets this line as to the effect that the *conjunction* of all of these *kath' hauta* terms converts with the subject.

210. Ross (1949), 582, following Jaeger (1960), 31-2, deletes *en*. The lemma has *huparkhei* for Ross's *enuparkhei*. As Mignucci (1975), 491 points out, the interpretation of Alexander, which Philoponus reports at 259,5-17, suggests that the reading of the lemma was that of Alexander, as well.

211. Ross, following Mure, deletes *en*, on which see the note on 84a16-17 on 582.

212. Ammonius.

213. See for example 1.19, 81b31, translated above as 'primarily.'

214. Ammonius understood *protôn* in the following lemma as having the sense 'first' while Philoponus understands it as having the sense 'prime'.

215. Ross puts the comma after *esti*, which changes the sense to 'but if this is so, number will inhere in it primarily'.

216. Reading *pleiosi* with G and a2.

217. We add a comma here.

218. Ross reads *en tôi heni* ('belonging in a single thing') while Philoponus reads *tôi heni*. Barnes, 180, argues that this allows Aristotle to escape the charges of offering an invalid argument, at the price of not presenting the required argument at all.

219. See n. 209.

220. i.e. taking a new term either above the major term or below the minor term, as opposed to in between both. This use of 'outside' is not the same as that of 268,2ff., according to which the term is applied only to terms of wider extent.

221. Philoponus' example does not apply to the way in which at 263,14 Philoponus describes the case under examination, since Socrates and Alcibiades, as particulars, are not the sort of item that is predicated of anything else, or which belongs to anything else. However, 264,3-4 makes clear that Philoponus is not restricting Aristotle's words at 84b4 'if one of them is not predicated of the other' to this sort of situation, for he there takes Aristotle to also have in mind two particulars, neither of which, in principle, can be predicated of the other, since particulars are not predicated of anything.

222. Accepting Wallies' insertion of *sômatos katêgoroito tou*.

223. i.e. mortal.

224. Philoponus reports that for *skhêma ti* at 84b8, some manuscripts read *trigônôi*.

225. Philoponus understands *ti* in the phrase *skhêma ti* at 84b8 to have the sense of 'a certain'; having angles equal to two right angles belongs to the subject not insofar as it is a shape but insofar as it is a triangle.

226. On this reading, the phrase at 84b8 would be translated 'insofar as it belongs to a shape'.

227. On Philoponus' lost work *Summikta Theôrêmata* see Verrycken (1990b), 253.

228. Reading *tou hupokeimenou* for *to hupokeimenon*.

229. Philoponus takes *atoma* to refer to individual subjects, in contrast to Ross (1949), 584, who takes the *atoma* to be immediate premises, and Mignucci (1975), 503 who takes them to be the last definienda.

230. Whole definitions are first in order of intelligibility. However, whole definitions do not feature among the immediate premises. Rather, definitions can be considered conjunctions of immediate predications (a conjunction of the predication of a genus of the definiendum and the predication of a differentia of the definiendum), and it is these immediate predications that serve as demonstrative premises.

231. Reading *deiknutai* for *deiknuntai*.

232. Omitting *kai to empsukhon logikon* with Wallies.

233. On the common notions see 254,30 and n. 187.

234. Ross reads *all' arkhê, kai* while the lemma has *all' arkhai kai*.

235. Philoponus is pointing out that the term 'principle' can be used in reference to an element of some composite, and the most basic combinations of those elements. He apparently understands Aristotle's assertion that there are as many principles as there are terms as speaking of principles in the first sense.

236. Ross has *hôsth'* for *hôste*.

237. At 84b33 Ross corrects A to D, but there is no evidence that this was the reading of Philoponus.

238. In the first figure, an affirmative conclusion requires an affirmative major premise, and a negative conclusion requires a negative major premise.

239. In Camestres the middle term is predicated affirmatively of the major term, but in Cesare it is predicated affirmatively on the minor term.

240. Only in the case of affirmative syllogisms does 'to fall outside' mean 'to be predicated of'. Because first figure syllogisms are such that the major premise and conclusion are correlatively affirmative or negative, negative first figure syllogisms will not be such as to have an affirmative major premise.

241. For the middle term to be of wider extent than an extreme is for it to be predicated of it, but in the third figure, the extreme is always predicated of the middle.

242. Ross, in contrast, deals with the problem by denying that Aristotle here addresses second figure syllogisms.

243. A 'thesis' (*problêma*) is a proposition put forward for proof; it is therefore offered as a conclusion of a projected syllogism. Within any syllogistic conclusion the minor term serves as subject. In the third figure, it has this role in both premises as well. Note that *problêma* at times refers to the question of which the thesis is the assertive version, in which case we translate 'problem'.

244. That is, that the minor term always serves as subject.

245. That is, the thickening of a syllogism comes down to the narrowing of the gap, so to speak, between middle terms, by adding further middle terms between them.

246. *An. Post.* 1.3, 72b23-5. The term rendered 'definitions' here is *horoi*, which can also have the sense of 'terms'.

247. On Philoponus' account, in the present context 'immediate' does not mean 'indemonstrable' but simply 'without a middle term' (regardless of whether or not a middle term could be supplied). In this context, the principle of a cognitive disposition is a faculty (such as intellect or opinion), while the principle of *any* syllogism is the immediate premise.

248. That is to say, within the first figure, the only negative premise in which the major term is involved is the denial of the major term of the middle.

249. Ross excises *ê mê panti*.

250. Ross reads *hôi dei* but Philoponus' commentary makes clear that he reads *hôi ou dei*.

251. The syllogistic form is Camestres.

252. See 268,22-8.

253. Ross has *badieitai* for *peseitai*.

254. This is discussed in 1.24.

255. This is discussed in 1.25.

256. This is discussed in 1.26.

257. 85a20-31.

258. 85a31-b1.

259. 'Itself in itself' translates *auto kath' hauto*, a phrase Plato uses in regard to separate subsistent Forms (see, for example, *Phaedo* 66A2 and 78D6. *Kath' hauto*, which we elsewhere translate as 'per se', here has the sense of 'by itself', a sense that Aristotle indicates at *Metaph.* 5.18, 1022a35-6.

260. 85b1-3.

261. Bracketing *hôste tôi men katholou* with Wallies.

262. Cf. 1.4, 73b32-74a3.

263. In the passage that Philoponus is paraphrasing (85b15-22) Aristotle says neither that a universal can be itself a substance, nor that it is *different* from that of which it is a universal. Philoponus might be here appealing to *Cat.* 5, 2a14-19, where Aristotle says that universal substantial items are themselves secondary substances.

264. Philoponus here attributes to Aristotle an account of universals according to which they are real, and have ontological standing apart from (*para*) particulars, but do not have subsistence (*huparxis*) outside of (*exô*) them. He does not make clear exactly what sort of ontological standing he takes Aristotle to grant them. For Neoplatonic accounts of universals, see Sorabji, *Philosophy of the Commentators,* vol. 3, 128-63; Sorabji (2006), (2010).

265. Their fault is not that they posit the universal as apart from (*para*) the particulars, as Philoponus would say that Aristotle agrees (273,14), but that they take it to be subsisting by itself.

266. Ross reads *anthrôpos* for the lemma's *ho anthrôpos* (see also 271,18); there is no difference in translation.

267. The subject of the clause could be either 'conclusion' or 'demonstration'.

268. Philoponus means 'the things that apply to isosceles as such'. If all triangles have a certain characteristic, it is no error to prove that all isosceles triangles have that characteristic, insofar as it belongs to the universal 'triangle'. It would however be an error to prove that characteristics that belong to isosceles triangles qua isosceles belong to isosceles triangles insofar as they are triangles.

269. Ross has *tautên* instead of *toiautên.*

270. The mathematical specialists prove that proportions alternate when the terms of the proportion are of this sort or that, but mathematics, in general, proves it when the terms are taken universally. This can be understood as proving it when the terms are, themselves, universals.

271. Ross punctuates as a question.

272. 275,15-18.

273. Mignucci (1975), 525-6 argues that Philoponus' proposed correction of Aristotle's wording is not necessary, since Aristotle's point is not that demonstration is more applicable to a universal than to the particular, but that, insofar as it is equally applicable to the particular, it is *not less* applicable to the particular.

274. It is not clear why *tên merikên* is in the feminine; one possibility is that it implicitly modifies *ousia* (substance).

275. This is an odd remark. Why would there not be a substance of a dog, which is referred to as 'dog'? An anonymous vetter proposes amending the text to *hei kuon kaleitai* ('insofar as it is called a dog'). The point would be that being called a dog is insufficient warrant for being a substance, as nonsubstances can also be called dogs (see LSJ s.v. VI and XII, according to which the term refers to certain moves or pieces in games); if a dog is indeed a substance it would be not on account of its name, but on account of its being an animal.

276. Bracketing *tis* with Wallies.

277. At 85b15-21 Aristotle says that the universal is, but he does not in 1.24 say that it is a substance.

278. *DA* 1.1, 402b5-8.

279. It is not clear to what discussion Philoponus is referring.

280. Omitting *ê ikhthun* with Wallies.

281. See *NE* 1.1-4.

282. Omitting *kai* with Wallies.

283. Philoponus does not mean that the subject of a per se predication is the cause of its own existence, but that it is the cause of the fact that the per se predicate belongs to it.

284. Ross has no comma before *hoti allo*.

285. To be constitutive of a species is not to be a formal principle or essence itself, as Philoponus is using the term in respect to attributes that follow from the form or essence of the subject. He is apparently here using the term to refer to what is a necessary and proper condition for belonging to a species.

286. As Barnes, 186, points out, Aristotle's argument, so understood, is fallacious. Particulars are indefinite *in number* (for, of any given kind, it is always in principle possible for there to be more of them) but that does not mean that they have no knowable determinate characteristics.

287. Philoponus is distinguishing the current sense of 'particular', as meaning 'less inclusive' from that which refers to the quantifier 'some', on which see *An. Pr.* 1.1, 24a18-19, 1.2, 25a1-25a13. He similarly distinguishes the current sense of 'universal', as meaning 'more inclusive' from the quantifier 'all', on which see *An. Pr.* 1.1, 24a18.

288. On this sense of formal (*logikos*), as applied to general arguments that apply to items of different kinds, see 234,4-235,7.

289. Following Wallies' emendation as indicated in the critical apparatus; the marking in the text is apparently a typographic error.

290. We amend *pasin* to *pasai*.

291. Because Philoponus starts a new sentence at 86b12 with *eti pros toutôi*, he must explain the anacolouthon.

292. 'Porism' is a term originating in mathematics, and is used to indicate a corollary or 'bonus' deduction from a previous demonstration.

293. We amend *pleon* to *pleona*.

294. Reading C, C, B, B with RUA2 for B, B, C, C.

295. Philoponus takes *tôn heterôn sullogismôn* at 86b23 to refer to second and third figure syllogisms. Against this see Ross (1949), 593 who argues that at 86b30-3 Aristotle considers only first figure syllogism, for which reason the phrase refers to supplemental syllogisms whose conclusions are the premises of the first figure syllogism in question.

296. Philoponus takes Aristotle at 86b30-1 to be saying that the principle (*arkhê*) of a syllogism is the major premise, as the quality (affirmative or negative) of the syllogism is determined by that. The 'inferior' premise is that which is less significant, for the minor premise does not serve as a principle in this way. In contrast, Ross, 593, takes Aristotle's meaning to be that the major premise is the principle in the sense that it expresses the conclusion in potentiality.

297. Philoponus here goes beyond Aristotle, who compares affirmation with being and negation with non-being, and on that basis states that the former is prior to the latter. Philoponus, instead, takes Aristotle to mean that specifically in regard to *substantial* being, being is prior to non-being.

298. Note that Philoponus does not use the Aristotelian *eis to adunaton* (except when quoting Aristotle), but instead chooses the expressions *dia tou adunatou* and *di'adunatou*, 'through the impossible', probably to emphasise the indirectness of the demonstration (*dia tou adunatou* is found in Aristotle (*An. Pr.* and *Top.*) but not in the *An. Post.*; *di'adunatou* is not Aristotelian). For the difference between *eis to adunaton* (*epagôgê*) and *dia tou adunatou* (usually *apodeixis/sullogismos*) see *An. Pr.* 1.44.

299. Accepting Wallies' deletion of *apophatikê*.

300. Accepting Wallies' insertion of *ê*.

301. Adopting a's reading: transposing *ek pollou* and reading *dêlon hoti ek pollou tou periontos* instead of *ek pollou dêlon hoti tou periontos*; the fact that a has *hê kataphatikê tês apophatikês* instead of the inverse, does not change the meaning of the sentence.

302. See 291,17-293,30. The relevant chapters of *An. Pr.* are 2.11-13 on demonstration to the impossible in the three figures, and 2.14 on the differences between probative demonstration and demonstration to the impossible.

303. See 293,31-294,12.

304. See 294,12-24.

305. See 294,25-295,25.

306. This direct syllogism is the example Aristotle gives at 87a3-4. He used the same premises at 86b16-17, where he was discussing privative syllogisms, which explains why the example is negative (the syllogistic form is Celarent).

307. i.e. *An. Pr.* 1.4-6.

308. The main source of Philoponus' discussion of demonstration to the impossible is *An. Pr.* 1.23.

309. The examples are given at 292,16ff.

310. i.e. either because of an invalid figure, or because of false premises.

311. In general a prodeduction is a deduction the conclusion of which forms the major premise of the next deduction. See e.g. Philop. *in An. Pr.* 263,5-6 and 264,4ff., ad *An. Pr.* 42b5. In the current passage, however, the prodeduction provides the first premise of the hypothetical deduction (namely 'if A belongs to some C, then D belongs to some C').

312. Accepting Wallies' deletion of *adunaton*. The hypothetical syllogism proves, not the impossibility of the assumed minor premise of the categorical syllogism (i.e. 'A belongs to some C'), but proves, through the impossibility of that premise, that its converse (i.e. 'A belongs to no C') must be true. This is clear from the subsequent analysis of the hypothetical syllogism, as well as from the example. In both cases, the result of the hypothetical syllogism is a universal negative conclusion.

313. Thus the whole syllogism through the impossible can be reconstructed as follows: *Categorical prodeduction*: [D belongs to every A] (assumption); A belongs to some of the Cs (assumption, opposite of conclusion to be demonstrated); therefore D belongs to some of the Cs. *Hypothetical, modus tollens*: If A belongs to some of the Cs [and D belongs to every A], then D belongs to some of the Cs; but it is not true that D belongs to some of the Cs; [and D belongs to every A]; therefore it is not true that A belongs to some of the Cs; therefore A belongs to none of the Cs.

314. Not accepting Wallies' addition of *on* at 292,21.

315. If we want the conclusion of the syllogism through the impossible to be universal negative, the assumed premise which leads to an impossible conclusion has to be its converse, i.e. particular affirmative. Since Philoponus states that that particular affirmative premise can only be the minor premise, he must be thinking of the first and second figure, as in the third figure a particular affirmative premise does occur as major premise in Disamis.

316. 'If, on the one hand' translates *ei men*. The subsequent 'If, on the other hand' (*ei de*) follows at 293,3.

317. Syllogisms with a universal negative major premise and a universal affirmative minor premise are Celarent (first figure) and Cesare (second figure); those with a universal negative major premise and a particular affirmative minor premise are Ferio (first figure), Festino (second figure) and Ferison (third figure). There is, however, a syllogism with a universal negative minor premise (Camestres, first figure), which Philoponus seems to forget.

318. There is no mood in the first figure with a particular major premise or a negative minor premise. The two moods which have a particular negative premise are Baroco (second figure, minor premise) and Bocardo (third figure, major premise).

319. This is a non-Aristotelian sense of 'circular demonstration' (for Aristotle's description of circular demonstration see *An. Pr.* 2.5). What Philoponus means here is that demonstration through the impossible reaches its conclusion indirectly, and by returning to its starting point. This idea is also found in Ammonius *in An. Pr.* 69,26-8; cf. Philoponus *in An. Pr.* 90,10-11, 248,3-5.

320. This seems to be a nod to the ongoing controversy between Stoics and Aristotelians concerning which logic has priority: propositional logic or the categorical syllogism. On this topic see Frede (1974), 114-15. Although Philoponus would maintain that direct demonstration is a categorical syllogism, he is here conceding that one might well make a case that a direct demonstration is embedded within a propositional syllogism.

321. esp. *An. Pr.* 1.23.

322. For an overview of modern objections see Barnes, 180-1, and Detel, 450-8.

323. Accepting Wallies' addition of *to B tôi C panti*.

324. Especially in Chapters 1.4-6.

325. i.e. to show that it is possible to reach one and the same conclusion ('A belongs to every C' in the example), using the same terms, both through direct deduction and through a deduction through the impossible. Note that Philoponus here changes the order with respect to Aristotle's example at 87a3ff. and gives a deduction through the impossible 'with regard to the conclusion itself', in response to the criticism mentioned at 294,12-14.

326. cf. above 294,2ff.

327. For a different solution see Barnes, 180, who proposes that Aristotle in the *An. Post.* is mainly interested in universal propositions.

328. Translating 'simple propositions', despite the fact that Philoponus says *huparkhousas*, rather than *haplôs huparkhousas*. Cf. however 296,10. For Philoponus' reference to Aristotle, Wallies suggests *Int.* 7, 17b29ff., which is probably right, although that is not immediately clear. The reference to the *Int.* in the context of the equivalence of indeterminate (*adioristai* (Aristotle) or *aprosdioristai* (commentators)) statements to particular ones seems to be a topic among commentators (e.g. Them. *in An. Pr.* 97,14-15; Amm. *in Int.* 111,31-2; Philop. *in An. Pr.* 79,4-5; Anon. *in Int.* 53,7-8; cf. Crivelli 244). However, the closest Aristotle comes to saying anything about such an equivalence is his statement that the indeterminate statement is *not* equivalent to the universal one. The term *isodunamein* is not found anywhere in Aristotle. See also below 296,10.

329. cf. 163,18.

330. As opposed to the *ad impossibile*, which starts from premises that are agreed to be false and hence leads to a positive conclusion indirectly. See *An. Pr.* 2.14, 62b29-31.

331. cf. above 295,16 and note.

332. On when this is possible, see *An. Pr.* 2.2.

333. Philoponus' point here is that in the case of a syllogism through the impossible, the conclusion reached by the syllogism is known to be false, and hence the syllogism cannot be sound. Sometimes a true conclusion results from false premises. To exclude such cases, Aristotle adds that 'this (i.e. the conclusion) is known and agreed to be impossible', in other words, we know it to be false. Philoponus says that in the case of a true conclusion resulting from false premises, it is the choice of extreme terms which is responsible for the truth of the conclusion, not the premises, i.e. not their connection to a middle term. The example Philoponus gives at 312,13ff. is 'human being is a stone, stone is an animal, therefore human being is an animal'. The conclusion is true because of the extreme terms 'human being' and 'animal', but not because of (their connection to the middle term 'stone' in) the premises, which are both false.

334. The word *sterêtikê* is not included in the lemma.

335. Although Aristotle speaks of privative propositions, he does not say 'that A does not belong to C', probably because the conclusion of the demonstration to the impossible is that A does belong to C, which, however, is impossible. We have added the negation, because Philoponus assumes that it is there ('the negative conclusion').

336. Both 'premise' and 'propositions' translate *protasis/-eis*. Philoponus uses the word *protasis* both for the premise and for premise and conclusion taken together, as does Aristotle at 87a13.

337. On Philoponus' view on these 'tekmeriodic' proofs see Morrison (1997), with the criticism of De Haas (1999), and Sorabji (2005), vol. 3, 265-8.

338. Natural demonstration (*he kata phusin apodeixis*) is not a technical term. Philoponus uses the term, which echoes Aristotle's 'by nature' (*phusei,* 87a17), for proper demonstration 'of the reason why' as discussed in *An. Post.* 1.13.

339. Only the part of this statement that is quoted from the lemma is in Aristotle. Cf. *An. Pr.* 1.29, 45a34-6, where Aristotle mentions that proof *per impossibile* is from consequents and antecedents of the terms in question.

340. Keeping the *oude* Wallies omits. The argument is a bit unclear but the general sense seems to be the following: the antecedent of a hypothetical syllogism is not a minor or major syllogistic premise – so, a fortiori, it is not a syllogistic premise at all. Nor is it one of a string of syllogisms. Now if demonstration is syllogistic, and these premises are not in any way part of a proper syllogism, they are not part of a proper demonstration. Note that the *affirming* of the antecedent can, as Philoponus himself also shows at *in An. Post.* 44,13-25, very well be a (minor) premise in modus ponens. Cf. 'Ammonius', *in An. Pr.* 68,18-23.

341. On the ancient position that there cannot be a syllogism from one premise, see Frede (1974), 20-1.

342. The major premise contains the predicate term of the conclusion, whereas the minor term contains the subject term of the conclusion. The major term holds of the minor term, which means, in terms of sets, that the minor term is contained in the major term. Thus the major is a whole of which the minor is a part.

343. Accepting Wallies' proposal to read *en tais apophasesin* where the text has *en tautais apophasis* (298,14-15).

344. Tentatively accepting Wallies' proposal. There is a textual problem. The text reads *enioi de katholikôteron * * * tês elattonos, en hêi kai ho meizôn merikôteros estin* (the lacuna is introduced by Wallies). Wallies seems to follow MS a1 in reading *katholikôteran* and then to supply *tên meizona phasi* for the lacuna. If he is correct, and assuming that 'more universal' and 'more particular' here are equivalent to 'more extensive' and 'less extensive' in the foregoing, Philoponus is referring to the following: in a universal negative major premise, the middle term and the major term are called 'coextensive', because the premise converts. Whether the terms are actually coextensive – as e.g. human being and capable of laughter are – or not, is irrelevant, since the premise is universal and negative and hence the two terms are like disjoint sets. Therefore, and here we reach the sentence with the lacuna, according to some people, the actual extension of the major term in the conclusion of a syllogism with a negative major premise (it seems we are always speaking of the first figure here) is irrelevant. Even if we take a 'more particular major term' – which we take to mean 'a major term with a smaller range', say on the level of species rather than genus – the major premise is still more universal, in the sense of being more extensive, because of its role in the syllogism.

345. Ross reads BC for AB of the lemma. Barnes reads BC ... AB.

346. The negations in question are 'A does not belong to B' and 'A does not belong to C' (*An. Post.* 87a14 as Philoponus reads it).

347. Philoponus here reformulates the conditional subclause of the beginning of the sentence, to pick up his train of thought after the long parenthesis.

348. As on other occasions, Philoponus uses the term 'theorem' here almost in its technical sense, a proposition to be proven. See above n. 2.

349. On precision (*akribeia*) in Aristotle see Barnes, 189-90, and Richardson Lear (2004), 103-4, 109-15. Philoponus distils three rules for determining the precision of a science from *An. Post.* 1.27: the number of items on which it depends (87a34-5), whether it is said of an underlying subject or not (87a33-4), and whether it is at the same time of the fact and of the reason why and not of the fact separately from the reason why (87a31-3).

350. Accepting Wallies' deletion of *hoti*.

351. That is, whenever the moon's position is perpendicular to the earth's surface from the point of view of the observer. We choose to translate 'standing before it' rather than 'covering it', as the latter would make the subclause superfluous.

352. Since the latter science proves a fact (the shape of the moon) on the basis of its effect (the appearance of the crescent), it is less precise than the former, which explains the eclipse on the basis of its cause. On this distinction cf. *An. Post.* 1.13.

353. The Greek has a feminine plural 'the first know' (*hai ... proterai ... isasi*), which refers to the first two examples (regarding sun and moon respectively), apparently considered as two instances of *epistêmê*.

354. Accepting Wallies' insertion of *to*, but rejecting his deletion of *sphairoeidês*.

355. This addition seems to pick up the phrase in the lemma 'and prior to it', and explains the priority of a science in terms of it starting from what is prior in the order of things.

356. As Ross, 596, points out, Philoponus needs to invert the order of the text to get the correct meaning. This is because Aristotle would not accept the thesis that the science that studies both the fact and the reason why is superior to that which studies only the reason why.

357. cf. the different levels of precision of ethics and mathematics as discussed at *EN* 1.3, 1094b11ff.

358. We translate 'not a positing' here for *ou thesis*, because Philoponus seems to pun on the being *a-theton*.

359. cf. Philop. *in An. Pr.* 97.11-12; Alex. *in An. Pr.* 86,5-6. There is no evidence that the Pythagoreans used the term *ousia* in respect to mathematical entities. Aristotle does tell us in *Metaph.* 1.5, 987a13-19 that for the Pythagoreans infinity and unity are the *ousia* of all things. Zabarella (983D), Ross, 596-7, and Mignucci, 573, rather suggest that *ousia* here has the sense of last subject. On the symbolic interpretation of Pythagorean number theory, see O'Meara (1989), 21 n. 46, 99, 148, 198.

360. Philoponus takes the unstated antecedent of *hosa* at 87a38 to be *theôrêmata*, but the term more likely refers to all of the entities that make up the genus studied by a science. See Ross, 597.

361. From the examples given it is clear that 'principles' (Aristotle's 'first things') refers not to propositions but to simple entities (or possibly concepts), that can function as subjects of theorems and that are not composite, as opposed to the ones that are composed from them. It is not entirely clear whether Philoponus – or anyone else in the Aristotelian tradition – takes theorems to be composed of simple entities or (mental) concepts. Both are problematic. Goldin (2010), who discusses two traditions of interpreting Aristotle's views on definition and demonstration in *An. Post.* 2.1-10, shows that Philoponus(?)' reading *in An. Post.* 2 suggests that the tradition which takes definitions to concern different kinds of *predications* starts with Philoponus(?) or Ammonius, whereas the tradition which takes them to concern different kinds of *entities* instead starts with Alexander. Philoponus here, however, seems to be closer to the latter. See also below, n. 399.

362. i.e. off a surface, for example.

363. Considering the following sentence, 'genus of theorems' probably means nothing other than that the theorems concern one scientific genus.

364. Note that five lines earlier Philoponus explained 'of them' as 'of principles'.

365. The reference is to Euclid *Elements* prop. I 5: 'In isosceles triangles the angles at the base are equal to one another, and, if the equal straight lines be produced further, the angles under the base will be equal to one another.'

366. 'Attribute of a theorem' is thus explained as 'attribute of a principle expressed in a theorem'.

367. This probably refers to 302,24-5.

368. This is a surprising formulation. Philoponus here gives some examples of 'attributes of them', i.e. the principles. We saw earlier that 'principles' refers to concepts not to propositions. However, whereas above Philoponus mentioned as attributes of principles 'things such as bending, being deflected, meeting and the like', he here introduces two examples as 'the assumption' of certain principles (straight lines and angles) having certain attributes (bending and 'such and such').

369. Reading *haterai* or *hai heterai* at 303,5; see Ross, 597.

370. That is, if one science uses principles that are the theorems of another, it cannot strictly speaking be called a different science. Philoponus here goes beyond Aristotle, who describes them as separate sciences, albeit with a genus that is the same 'in some sense' (*An. Post.* 1.7, 75b9; 1.13, 78b34ff.).

371. That is, Philoponus takes Aristotle to present evidence of the fact that 'the [undemonstrated principles] have to be in the same genus as the things that have been demonstrated' (87b1-3), rather than a second piece of evidence in support of 87a39-b1.

372. cf. *Phys.* 185a4-5.

373. i.e. the relation of principle to that of which it is a principle. This idea is not explicit in Aristotle, but is common in the later commentary tradition. Cf. Plotinus, *Enn.* 6.3 28,4-8.

374. 'The same series of predications' seems to refer to a series or chain of middle terms (C, D, and F as middle terms for AaB), in which the lower (or as Philoponus will call them 'subordinate') middle terms are species of the higher ones. Such middle terms are non-continuous if they are not immediate.

375. *Lemmatia*. In his commentary on Euclid's *Elements* (211,1-212,4), Proclus distinguishes two kinds of lemmata: (1) any proposition which is assumed for the construction of something else; (2) a proposition requiring confirmation. The diminutive *lemmation* occurs once in Euclid (*Elem.* 10 prop. 41). In later mathematicians and in the commentators it is more common and seems to be equivalent to lemma, or refer to less important or less general assumptions (see e.g. Ptol.,

Syntaxis 43,2-5). As Heath points out (vol. I 33-4), among mathematicians lemmata became the later supplied proofs of implicit assumptions made by predecessors. Although it is not entirely clear what Philoponus has in mind here, this is probably what he is referring to. It is likely that different proofs were provided for each assumption. Moreover, Proclus (211,18ff.) tells us that the lemmata are found (which we take to mean that proof of them is found) using different methods: analysis, division, and *reductio ad absurdum*.

376. The question is whether the two terms from different series should be universally denied of one another.

377. Philoponus is thinking of Darapti. The point is the following: if two terms are not in the same series, and yet are capable of proving the same thing, they are at least somehow connected through that thing. That which is the subject term in both syllogisms, then, becomes the middle term in the syllogism which expresses their connection. In terms of sets, the subject term of both syllogisms is in the (non-empty) intersection of the two terms from different series.

378. Reading *ti* for *tí*, which makes no sense here. MS a has *hoti*, which does not attribute to Philoponus an assertion which is false or nonsensical, but does not provide us with the desired Darapti syllogism either.

379. Note that C and F are not used by Aristotle in the examples following, as he gives no further example of a demonstration of the same thing though different terms from the same series. Philoponus does not use C and F either, but he will give an example of two terms from the same series (moving and being altered, 305,9-12).

380. Considering the previous sentence, one would expect 'changing' here, not 'being moved'. However, as Mariska Leunissen pointed out, Philoponus' choice for 'being in motion' may be explained by the fact that he is discussing subspecies of motion. Thus the aim of the whole sentence is to show in what sense the claim 'it is possible to prove that what feels pleasure changes, both through the middle term "motion" and through the subspecies of motion' (305,8-9) is true.

381. This is the example Aristotle gives for G at 87b9.

382. Wallies suggests Philoponus has in mind 6.8, 238b25. Perhaps, however, 5.6, 230a4-5 is more suitable.

383. 'End' in the sense of *Phys.* 194a28ff.

384. Pleasure occurs especially in being brought to rest, because 'pleasure is in rest, rather than in motion' (*EN* 7.14, 1154b27-8). By adding the clause 'in all those etc.' Philoponus excludes kinds of motions which can come to a standstill without this bringing a feeling of pleasure to the substances undergoing them. The addition may be inspired by the example given by Aristotle just before *Phys* 5.6, 230a4-5 (possibly referred to by Philoponus, cf. n. 382 above): the motion 'from health to disease' (230a3). Note that, by subsuming 'coming to rest' under 'motion', Philoponus is (inadvertently) disqualifying Aristotle's example, since 'coming to rest' was presented as a middle term from a *different* series of predication. On the other hand, Philoponus here solves a controversy between Plato and Aristotle, of which Aristotle's example is part: according to Plato, pleasure was motion, whereas according to Aristotle, it was coming to rest (*EN* 7.14; 10.3-5). By qualifying 'coming to rest' as intermediate between motion and rest, Philoponus harmonises these two views.

385. Considering the 'one another', Philoponus should have said here 'and moving holds of something that is being brought to rest'. Instead, as the text stands now, it merely reformulates the very point made in the previous phrase, 'for being brought to rest belongs to something that moves'.

386. This is clear because in the other figures only the position of the middle term changes, not its function. To give an example, in Cesare one could prove that 'no human being is a stone' through middle terms from different series of predication: no stone is an animal, all human beings are animals, no human being is a stone; no stone is capable of laughter, all human beings are capable of laughter, no human being is a stone.

387. There is a problem here: there is no requirement that the premises of a deduction be necessary or for the most part. As Mignucci, 596, points out, at *An. Pr.* 1.13, 32b21-2, Aristotle asserts the contrary.

388. For Aristotle's definition of chance (*tukhê*) see *Phys.* 2.4-6. In Chapter 5 Aristotle points out that what happens by chance happens neither always nor for the most part. At *Top.* 112b10-11 Aristotle says that 'in very few cases' (*ep' elatton*) is called the opposite of 'for the most part' (*hôs epi to polu*).

389. The 'someone' might refer to people holding the 'Heraclitean' and 'Protagorean' views expressed and refuted in Plato, *Theaet.* 151E-187A; for Aristotle on the role of perception in scientific understanding see *An. Post.* 2.19 and *Metaph.* 1.1.

390. Considering these examples, we should probably take 'quality' (*poion*) here in a broad sense: strictly speaking, geometrical figures belong to the category of quantity.

391. On the common sensibles see *DA* 3.1.

392. cf. *An. Post.* 87b29-30.

393. The customary way of describing the universal in Aristotle is as holding 'always and in every case'. In *An. Post.* 2.12, he describes particular events which 'come about ... have come about and will be' as opposed to things that come about universally, 'always and in every case'. Philoponus seems to combine the two phrases.

394. Philoponus is being very sloppy here: he gives a plural *diaphorai*, 'differentiae', which would have to be the subject, except the verb is a singular (*aphorizei*, 'determines'). As Wallies points out, in the *Categories* (not the *Isagoge*), we find a passage that is very similar to what Philoponus says here, and which contains the singular verb but with a singular subject: *to de eidos kai to genos peri ousian to poion aphorizei* (3b19-20; cf. Porph. *in Cat.* 96; Alex. *in Metaph.* 236,6, 496,38). Philoponus may be confusing the *Cat.* passage with a passage in Alexander: (*in Top.* 314,23: *hê gar en ousiai diaphora peri ousian men to poion aphorizei*). The confusion may have been caused by *Metaph.* 1020a33 (*to poion legetai hena men tropon hê diaphora tês ousias*). Elsewhere, e.g. at *in Cat.* 73,17 Philoponus does refer and quote correctly.

395. Although Aristotle does not need to show that perception is not scientific understanding, he does so in order to present the complete range of cognitive faculties.

396. Depending on the reading of the MSS, Aristotle says either that 'scientific understanding' *is* 'getting to know the universals' (*hê de epistêmê to to katholou ginôskein estin*, 87b38-9, as in Ross) or that the former is *by*, or comes about by, the latter (*hê de epistêmê tôi to katholou ginôskein estin*, as in Barnes). Philoponus chooses the latter, despite the unlikely construction, thereby expressing a causal relation, rather than stating an identity. See also n. 403.

397. Unqualified 'intellection' (*noêsis*) here is an immediate grasping without 'the reason why', which will later be divided into the grasping of first principles, the cause of which cannot be known because they have no cause, and the grasping of everything else, which can be known through its cause, and in fact is better known through its cause than without it. In the *Posterior Analytics, noêsis* generally concerns knowledge of universals (McKirahan [1992], 258-9). In Neoplatonic epistemology, *noêsis* is the cognition which concerns the intelligible.

398. 88a7, discussed at 310,17ff. Cf. *An. Post.* 2.8-9, where Aristotle mentions the class of things that do not have a different cause. Philoponus(?) in his commentary on these chapters understands it, not as a class of things, but as a class of predications (364,1ff.). On this passage see Goldin (2010), esp. 180. See also above, n. 363.

399. Accepting *ei* and taking *ekeina* to refer to implicit 'others' than *toutôn* in line 13. 'The definition of intellection ... is different', i.e. we use a different concept of intellection.

400. On the common notions see 254,30 and n. 188.

401. The issue is raised at 88a11 (311,1ff.). 'What we meant earlier' refers to *An. Post.* 1.18 (213,15ff.).

402. 'Deduce' (*sullogizesthai*) must be used here in a broad sense of 'derive', as in fact intellect does not use the method of deduction. If anything, it *induces* them from particulars (cf. *An. Post.* 2.19).

403. cf. 87b38-9. It is surprising that Philoponus calls the 'deduction' of universals 'scientific understanding'. Scientific deduction (in Barbara, with the cause as middle term) leads to scientific understanding, but the grasping of universals is superior to it. In his earlier quotation of 87b38-9 (which is not quoted in the lemma) as expressing the causal relation that 'scientific understanding comes about by getting to know the universals' (307,27-8 and n. 396), Philoponus seems to be aware of this. Here at 309,2, however, Philoponus seems to follow the reading preferred by Ross, i.e. the identity statement that 'scientific understanding is getting to know the universals'.

404. Philoponus seems to combine two examples from *An. Post.* here: at 88a14-15 Aristotle mentions the puzzle of light moving through glass; at 2.11, 94b28 he uses an example involving lanterns. In Aristotle's time, lanterns were made of horn or skin. In Philoponus', apparently, they were made of glass.

405. This is the theory of light underlying the example at *An. Post.* 2.11, 94b27-31. On different theories of light in the commentators see Sorabji (2005), vol. 2, 274ff.

406. That is, the result would not be that the air would be illuminated as a continuum, without interruptions. Instead, some parts would, and some parts would not be illuminated.

407. This theory of light is also found in Aristotle, *DA* 2.7, 418b9-10. Cf. Philoponus' comments at *in DA* 324ff., esp. 329,3ff. Light, according to Aristotle, is the activity of what is transparent *qua* transparent. Philoponus' 'the transparent body' probably refers to the material of which the lantern is made, and which is transparent to a certain degree. On degrees of transparency in bodies, see Alexander *DA* 455ff.

408. On Philoponus' reading of this sentence, see nn. 396 and 403.

409. cf. 307,28-34. 'Of perception' translates *tês aisthêseôs*. The genitive could qualify either *logôi* or *to katholou* (the universal of perception). Although *tou katholou aisthêsis* (310,12) and *tês aisthêseôs to katholou* (13) form a nice chiasm, we have chosen to understand *tês aisthêseôs* as qualifying *logôi*, which is more informative. Philoponus may just be referring to reason's using perception as a tool (cf. 307,31-3; *in DA* 403,29), but a somewhat more daring interpretation could be that he refers to a specific kind of reason, namely the one that is involved with perceptibles. Cf. Alcinous' 'opiniative reason' (*Didask.* 4, 154,25-9). See also Sorabji (2005), vol. 1, 1a.

410. See Philoponus(?) *in An. Post.* 342,21, tr. Goldin with n. 71.

411. We agree with Wallies' careful suggestion ('deleverim') to delete *de*.

412. Ross has *hualon* for *huelon* of the lemma.

413. Accepting Wallies' suggestion *elenkhei*.

414. The verb here, *kaiei*, which is well attested in the manuscripts of the *An. Post.*, is not apposite, as the glass itself does not burn (unless Aristotle is thinking of glass lenses which can be used to ignite something). Another reading (found in two manuscripts) is *kai ei*. As Mignucci, 609, points out, Philoponus takes *kaiei* to be synonymous with *phainei*, a distortion of the sense of the term.

415. *Dia to horan*: Aristotle does not, as Wallies says, leave out *dia*, but has a dative *tôi*.

416. 'Understand' translates *noêsai*, the aorist infinitive of *noein*. We take it to express in a quite general sense the result of thinking. 'Taken together' translates *hama*, which Philoponus takes to contrast with *khôris*. It is not entirely clear whether *khôris* and *hama* modify the verbs or the 'cases'. Probably a bit of both. Since in Philoponus' interpretation 'into a unity' corresponds to *hama*, he apparently understands it to modify the 'cases' (*pace* Barnes, 185-6).

417. The proximate premises are those from which a conclusion is derived (cf. 313,30ff.). These premises are moreover proper to the genus in question (ibid.), as opposed to the common principles (cf. 315,27ff.). For this distinction see *An. Post.* 1.10.

418. What Philoponus should have said here is that at least one of the premises has to be false.

419. This is neither true (as Philoponus himself points out) nor is it what Aristotle says. Aristotle does say, in *An. Pr. 2.2*, that it is possible to draw a true conclusion from false premises, but that this conclusion is true only with respect to the fact, not to the reason why (53b4-10).

420. 88a21.

421. This 'thickening' is used by Aristotle in the sense of making deductions (that are not of the first figure) more universal (79a30) and of increasing the number of middle terms 'until they come to be indivisible and unitary' (84b35). Philoponus is here thinking of the second sense, which he explains as thickening 'down to the point at which you arrive at immediate premises, which are principles of the syllogisms and are as it were certain indivisible unities' (269,5ff.).

422. The point being that, if you would go back from the conclusions to the premises, whatever they are, in the case of false conclusions the premises are always false, and in case of true premises the conclusions are always true. In the case of a true conclusion from false premises, the premises of those premises are themselves necessarily false and the true conclusion itself will be a premise only of true conclusions. That is, in a series there will be only one case of a true conclusion from false premises.

423. As at 312,10, Philoponus seems to falsely assume that deductions which have false conclusions also have two false premises.

424. As is shown by the example. Also the parenthetical remark at the end of the note probably should be its own sentence.

425. Note that 'they can both hold' does not mean that they can hold *at the same time*. As is clear from the example, Philoponus is speaking of contingent falsities, the negations of which are contingently true.

426. A 'contrary falsehood' (*pseudos enantios*) is a falsehood which predicates something of its contrary.

427. i.e. that 'human being' and 'horse' coexist in one subject, or 'stone' and 'stick'. Philoponus here makes Aristotle's discussion more precise. See also the discussion of the next lemma.

428. Philoponus gives *heautois* the sense of *allêlois*.

429. 'The equal is more' and 'the equal is less' are both falsehoods, and both for two reasons: (1) 'the equal' can never be either 'more' or 'less', and hence both are *impossible* falsehoods comparable to 'justice is cowardice'; (2) on the other hand, 'more' and 'less' are forms of inequality, and are in that sense the *contrary* of 'equal'. Note that the statement regarding 'the equal' at *An. Post.* 88a27-30 is slightly different from Aristotle's discussion of that same notion at *Metaph.* 10.5, where he points out that strictly speaking 'unequal' is not the contrary of 'equal'. Philoponus' addition of 'equal simpliciter' is perhaps intended to exclude cases such as those mentioned by Aristotle at *EN* 1108b15-16, 'equal compared to more' and 'equal compared to less', as in these cases 'the equal is less' and the equal is more' respectively would in fact be true.

430. Following Philoponus' interpretation, the argumentation adduced in defence of the thesis that not even the principles of all truths are the same can be summarised as follows. There are two sub-theses, namely (1) the proximate principles cannot be the same; (2) the first principles cannot be the same (313,28-30). The first thesis is refuted with one argument, the second with eight. Argument against (1): premises have to be akin to the conclusions; the conclusions of different sciences concern different genera; therefore the premises have to concern different genera (313,31-315,24). Against (2a): the first or common principles do not suffice for demonstrations: a genus-specific premise is always required (315,25-316,22); (2b): for us to know that the principles are the same, they have to be finite; the conclusions are infinite; the number of premises is almost equal to the number of conclusions; so the number of premises is infinite and the principles cannot (be known to be) the same for all sciences (316,23-318,18); (2c): true conclusions can be divided into necessary ones and ones that can be otherwise; they have necessary principles and principles that can be otherwise respectively; hence the principles cannot be the same for all sciences (318,19-26); (2d) understanding 'the principles are the same' as the principles of one science being the same as those very principles is absurd (318,27-319,5); (2e): we should not understand 'the principles are the same' as (a) all things can be proved from any given thing; (b) each thing can be proved from all things, on the basis of evidence and argument (319,6-10); (i) evidence: in mathematics we do not prove everything from the same principle (319,10-17); (ii) argument: analysis of different conclusions leads to different immediate premises (or definitions) from which they are derived (319,17-31); (2f): if 'principles' is understood as 'immediate premises', there is (at least?) one for each genus; genera are different; therefore the principles are different (320,3-15); (2g): the option that all principles are of the same genus is rejected for the reason expressed in 1a (320,16-321,6). After this argument, and before concluding that it is in no way possible for the principles to be the same for everything, Philoponus briefly points to a way in which they are the same. See 321,6-16 and note.

431. See *An. Post.* 88a18-19.

432. i.e. deductions in *Barbara* (universal affirmative premises and conclusion), with true premises.

433. While Philoponus takes *ta keimena* to refer to the deductions presently under consideration, Ross, 601-2, takes the phrase to refer to points concerning demonstrative science that have already been established.

434. The proximate principles are those premises from which conclusions are actually derived, as is shown by 314,1-2. They are genus-specific. The first principles are the immediate premises, common notions, and definitions (cf. 315,27-9).

435. For this argument and the examples see *An. Post.* 1.7.

436. This is an interesting example, as Aristotle in his discussion of scientific kinds in *An. Post.* 1.7 mentions harmonics as a science subordinate to arithmetic (75b16) – and Philoponus seems to agree with this in his comments (17,14ff.) – and hence as concerning (discrete) quantity, rather than quality (cf. *in An. Post.* 146,6ff.; *in DA* 375,32-3). Philoponus is here clearly thinking of a science of music as concerning perceptible qualities, namely sounds (cf. 306,20-5). Perhaps for reasons unknown Philoponus here agrees with Theophrastus, according to whom high and low pitch are qualities, not quantities (on this topic see Barker [2007], ch. 15).

437. Note that, to explain the difference in genus, Philoponus focuses on the different categories the subject matters of the disciplines in question belong to.

438. 'Akin to' translates *suggenês*, the root meaning of which is 'sharing a genus'.

439. For the view that the point is a unit with the addition of a position cf. *An. Post.* 87a35 and 301,8ff. Note that according to the Neoplatonic view of unit and monad, this is the reason why geometry is subordinate to arithmetic, as the point of geometry is dependent on the unit of arithmetic.

440. The definition of a line as 'a magnitude in one dimension' (*megethos ... eph'hen diastaton*) is cited by Proclus (*in Eucl.* 97,7-8) as an alternative to Euclid's 'negative' definition of the line as 'a breadthless length'. Although the alternative definition is not as such in Aristotle, Heath does take its source to lie in *DC* and *Metaph.* (for references see Heath, 158-9).

441. Ross has *harmottein* for *epharmottein* of the lemma.

442. 'In the middle' (*eis mesa*): This phrase is ambiguous between 'in the middle' and 'at the middle terms' (i.e. having the position of the middle term). However, considering the fact that 'from above' and 'from below' are explained by Philoponus as also referring to the place of the middle term, albeit in the second and third figure respectively, we have to translate 'in the middle'.

443. It is not entirely clear what the subject is, here, as in the previous sentence Philoponus switched from 'the principles' (which were the subject at 88a33) to a vague neuter. Further on, he will follow Aristotle (88a35) and switch to 'terms' (*horoi*).

444. 'Of the second figure' translates Philoponus' *tou mesou*, literally 'of the middle (*sc.* figure)'.

445. Note that in all three figures the terms in question are effectively the middle term.

446. 'In the same way' (*houtô*) should not be taken too strictly. Wallies rightly refers to Aristotle's discussion of the terminology 'middle term' and 'extreme' at 25b36, 26b39 (second figure), and 28a15 (third figure). The first figure middle term is described there as 'coming in the middle position' (25b36). Aristotle does not, however, use 'from above' or 'from below' for the second and third. Instead, we find 'standing outside the extremes and first in position' (26b39, cf. 'above') and 'standing outside the extremes and last in position' (28a15, cf. 'below').

447. e.g. Alexander at *in An. Pr.* 81,28ff.

448. e.g. dogs are animals, horses are animals, therefore dogs are horses. The premises are true (i.e. the predicate holds of the subject), but the syllogism is invalid. Philoponus need not specify whether the premises are universal or not, since in the second figure all moods contain at least one negative premise.

449. The question is not 'what does "from above" mean?', as he just explained that, but rather 'how can he speak of applying if one of the premises has to be negative?'

450. That is, we are speaking on the level of premises, not of syllogisms. 'Combination' translates *sumplokê*, a word which Plato (*Soph.* 259E6 etc.) and subsequently Aristotle (*Cat.* 1a16 etc.) use to indicate the combination of Forms/terms into an (atomic) sentence (*logos*). Among the commentators it is very commonly used to refer to the combination of premises to make a syllogism.

451. Accepting Wallies' insertion of *ê*.

452. Accepting Wallies' insertion of *prôtois*.

453. Outside in second and third figure: *An. Pr.* 26b39, 28a15; inside in first figure: 25b32-6.

454. This is a very convoluted way to say that, in universal affirmative predicates, by stating that a predicate holds of the subject, one ascribes the subject to a group that is possibly larger than the subject group. That is, the subject collection is an element of the predicate collection and is in that sense encompassed by the predicate.

455. Grammatically, 'some' seems to refer to 'kinds of beings'. Considering the word 'belong' (*huparkhei*), however, it seems reasonable to assume that Aristotle has in mind predicates (which, of course, also differ due to their falling under different kinds of beings). However, although this sentence is not quoted in the lemma, Philoponus' exegesis suggests that he has in mind both propositions (at 316,3-5) and subject or predicate terms (at 316,14-21), which is why we have kept the translation neutral.

456. On the common notions see 254,30 and n. 188. Since he goes on to call them 'common axioms', we can assume that Philoponus takes Aristotle to refer to the 'axioms' he introduces in *An. Post.* 1.2 (72a16-17), 10 (76a37ff., esp. 76b14, *koina axiômata*), 11 (77a22ff.). For the law of the excluded middle as a common principle see also *An. Post.* 1.11, 77a22-30. The law is also mentioned at *An. Post.* 1.1, 71a13-14; *An. Pr.* 32a27-8; *Top.* 143b15-16.

457. This is Euclid's first axiom (*koinê ennoia*).

458. Philoponus here mentions common principles in their function of premises, ignoring Aristotle's view of the analogical nature of axioms. As is clear from the lemma, according to Aristotle common principles are adapted to the relevant genus: 'things are proved through the common principles with these (i.e. genus-specific) [things]'. Cf. *An. Post.* 1.10, 76a38-b2 (on this topic see McKirahan, 68ff.). Philoponus instead thinks of two premises, one of which is a common notion and the other of which is a genus-specific premise. See also the introduction and cf. 316,14ff., where Philoponus does refer to the adapting of the common notions.

459. Philoponus here describes the role of the law of the excluded middle (LEM) in the demonstration through the impossible (cf. *An. Post.* 1.11, 77a22-4). Once it has been shown that the assumption that the diagonal is commensurable with the side leads to absurdity, LEM allows us to conclude that it is therefore incommensurable.

460. Accepting the emendation proposed by Wallies, and reading *ekeinai* for *monai*.

461. This reasoning is cogent only if one accepts the very Aristotelian assumption that each science concerns one genus or domain. Moreover, because he does not employ the analogical reading of common notions, Philoponus does not show 'that the universal and common notions are also not entirely the same for every science'. In fact, in his reading they are – they just do not suffice for demonstrations. Philoponus seems to be aware of this, as he adds 'which has the status of principle in relation to the conclusion', to emphasise the status of principle of the genus-specific premise in question. Moreover, he goes on to show that the common notions are not even applicable in their general form, but need to be adjusted to the subject matter at hand.

462. This probably refers to 10,27-11,3. See also the Introduction, n. 4.

463. Accepting Wallies' addition of *koinôs*. Here Philoponus does mention the analogical nature of common notions.

464. Euclid *Elements* 7 prop. 13: 'Let the four numbers *A, B, C,* and *D* be proportional, so that *A* is to *B* as *C* is to *D.* I say that they are also proportional alternately, so that *A* is to *C* as *B* is to *D.*' E.g. if 1:2::4:8, then 1:4::2:8.

465. i.e. not as a middle term between two terms of a premise, but 'outside', as Philoponus calls it (317,18ff.), and hence connected to one term (either the major or the minor term of the conclusion), not two.

466. For this distinction between proximate and first principles, cf. 276,1-6 and note.

467. On the relation between the number of terms, the number of propositions, and the number of conclusions in a syllogism, see *An. Pr.* 1.25, 42b1ff. Note, however, that according to Aristotle's analysis of the way the conclusions increase when a term is inserted in the middle Philoponus is wrong here: Aristotle makes clear that the added term will not only effect a deduction in relation to the two terms in between which it has been inserted (*An. Pr.* 1.25, 42b23-5). To use Philoponus' example, 'if C is posited as a middle term of "A B", two premises are generated, namely AC and CB, and one conclusion', namely AB. But 'if we wanted to prove the premise AC, by inserting as a middle term, for example F, then again two premises are added', *but not* 'one conclusion'. That is, we do prove the one conclusion AC, but since that conclusion is part of the syllogism AC, CB, AB, adding F will also add a conclusion FB.

468. Philoponus here gives two examples of conclusions doubling as premises, i.e. in syllogisms leading to the conclusion AD. In the first, the conclusion BD serves as a premise in a syllogism which concludes AD. In the second, AC, which was the original conclusion, and CD, which was the conclusion obtained by adding D, are both used as premises in another syllogism which concludes AD. We thank Richard McKirahan and Wouter Goris for their suggestions.

469. Richard McKirahan points out that after adding term D, we have three conclusions AC, AD and BD; after adding E we have six conclusions AC, AD, AE, BD, BE, CE, after adding F we have ten conclusions. We have therefore chosen to translate as we did, to express that the number of *new* conclusions is the same as the number of terms that is not adjacent to the added term (which makes the new premise together with the added term, i.e. 'the pair'). If we add one term, that leaves two terms non-adjacent to that term, which can form new conclusions, and hence we obtain a total of 3 (1 old + 2 new) conclusions. If we then add a second term, there are three non-adjacent terms and 3 old + 3 new conclusions. If we add a third term, 6 old + 4 new etc. Cf. Aristotle, who formulates it differently and says 'if one term is added, conclusions will be added less by one than the pre-existing terms' (*An. Pr.* 1.25, 42b18-19, tr. Jenkinson).

470. CD in the example.

471. As Mignucci, 634-5, points out, Philoponus' modal logic here is faulty, given that what is necessary is also possible.

472. Accepting Wallies' deletion of *to*.

473. Accepting Wallies' deletion of *tên*.

474. As Richard McKirahan pointed out to us, the division in question divides the genus, rather than a species of the genus.

475. As Ross points out ad loc., the commentary indicates that Philoponus reads *hou* for Ross's *ho* here.

476. Philoponus switches from propositions to terms here.

477. This is not entirely accurate, as only some common notions are used by all sciences (e.g. LEM), whereas e.g. the principle that 'if equals are subtracted from equals the remainders are equal' (*An. Post.* 1.10, 76a41) is limited to e.g. quantitative sciences. On the common notions in Philoponus see the Introduction, on the common notions in Aristotle see McKirahan (1992), ch. 6).

478. From here on 'apprehension' translates *hupolêpsis*, a wide term, defined by Aristotle at *DA* 3.3, 427b25 as covering *epistêmê, doxa* and *phronêsis*.

479. 89a11-13.

480. Accepting Wallies' deletion of *kai*.

481. cf. 89a11. This hypothetical reduction of opinion to knowledge relies on a strong assumption, namely that the differences between forms of cognition are actually determined by their subjects, so that identical subjects result in identical forms of cognition. This assumption is common in ancient thought, and can be traced to the formula 'like knows like'. Cf. Plato, *Republic* 477D; Aristotle, *DA* 2.4, 415a20-2. Interestingly, the second difference between knowledge and opinion (i.e. the mode of understanding, 89a16ff., cf. 322,5-8) is hereby reduced to a result of the first. For Aristotle on the ways in which there can be knowledge and opinion of the same thing, see 89a23ff.

482. The example of the necessity of fire's warming is given at *Int.* 13, 23a2-3. In his comments on that passage, Ammonius also gives the example of the necessity of heaven's circular motion (*in Int.* 240,7ff.). The latter is an example of eternal co-occurrence, the former of co-occurrence as long as the substrate exists.

483. This passage seems somewhat repetitive, but consists of three distinct steps: (1) knowledge and opinion differ in the mode of understanding (unchangeable or not); (2) this tells us how their subjects differ (necessary or not); (3) this tells us that knowledge and opinion are different in a second way, namely in having different subjects.

484. 'Things that are' translates *ta onta*. Considering the example, Philoponus seems to have in mind (true) sentences or propositions.

485. This seems to be a reference to 89b31-3.

486. Aristotle does not ask this question. He only asks how, if scientific understanding and opinion have the same subject, they would still be distinct. See above.

487. On knowledge as a standstill of the mind, see *An. Post.* 2.19, 99b36 of the perception (*aisthêma*), 100a6-b3 of the universal (*katholou*), and *DA* 407a33 of intellection (*noêsis*) and syllogism (*sullogismos*). Each of these is a standstill in its own way. A perception can stay in the mind in the form of a memory, the universal is a standstill with respect to the many individuals, intellection is non-discursive, and syllogism reaches a conclusion. The case of scientific understanding (*epistêmê*) is most similar to that of syllogism.

488. Philoponus is referring to the etymological relation between *epistêmê* and *epistasis* (standstill).

489. cf. *An. Post.* 1.15, 79b28, where Aristotle opposes simple conviction (*haplê hupolêpsis*) to conviction resulting from deduction.

490. 88b34-5 (Philoponus changes the word order).

491. Philoponus uses the formula known from mathematical proofs.

492. Actually, Philoponus is using *epistasthai* here, albeit not in the narrower *An. Post.* sense of 'to have scientific understanding', but in the very general sense of 'to know' or 'to understand'.

493. The terms used are *epistasthai*, *eidenai*, and *gignôskein*.

494. This is the traditional Platonic and Aristotelian epistemological notion of *doxa* (as compared to the ontological one, which centres on the kind of object it concerns).

495. That is not to say that every opinion, once provided with a deduction, becomes an opinion of the reason why, as there are opinions for which no deduction can be presented, e.g. those based on abductive reasoning, such as that there is fire, based on seeing smoke.

496. The passage from Alexander is not extant. On the non-existent goatstag as object of scientific understanding see *An. Pr.* 1.38, 49a22-4, *An. Post.* 2.7 92b4-8.

497. The reference is to Ammonius, whose lecture notes Philoponus used in compiling the commentary.

498. This is a difficult passage, but Philoponus is clearly adding one of his 'personal reflections', dissenting from his teacher Ammonius (cf. Sorabji [1987], 3-4). The argument can be constructed as follows: Ammonius may be wrong in rejecting the interpretation of Alexander. Alexander stated that truth is not limited to things that are, which is why here Aristotle adds 'and are the case', thereby limiting the extension of the things discussed. Ammonius apparently objected that such things are not the object of *doxa* – implying that what is not *cannot* be otherwise and hence need not be excluded in this passage, which concerns things that are but *can* also be otherwise. Philoponus tentatively suggests that Alexander's reading can be defended if we accept that also among what is not there are things that can be otherwise.

499. The *diakrisis* here cannot be a discrimination between two things since the candidates for discrimination would be true things on the one hand and things that can be otherwise but are not on the other – but Philoponus is arguing instead that being true, possibly being otherwise and not being can and do go together sometimes.

500. The idea that of our cognitive capacities the (human) intellect is the one reserved for knowledge of the divine is not Aristotelian. 'Simple intuitions' (*haplai epibolai*) is a Neoplatonic technical term with roots in Epicurean and Platonic epistemology and refers to immediate apprehension by the intellect. It is used especially for the intellect's direct apprehension (e.g. of the Forms, e.g. Proclus *in Tim.* 1, 438,28-439,1; cf. already Alexander *in Metaph.* 599,32 on intellect's simple apprehension of simple things). At *in An. Post.* 48,11-15 Philoponus says knowledge of the definitions is gained through intellect's *haplai epibolai*. Note that at 332,5ff. the part or virtue of intellect that is concerned with 'the divine and intelligible forms' is identified as wisdom (*sophia*).

501. The passage quoted is *An. Post.* 1.3, 72b23-5, the famous description of foundationalism. Differences between Aristotle and Philoponus' 'quote': (1) 'we say that there is a kind of' (*einai tina phamen*) in Aristotle comes after 'principle' (*arkhên*), so that the indefinite pronoun qualifies the beginning, rather than 'scientific understanding' (*epistêmê*); (2) Aristotle has 'we become acquainted with' (*gnôrizomen*) where Philoponus has 'we know' (*gignôskomen*), which allows a more Platonising/Christian reading of the knowledge of the terms. For Philoponus' interpretation of 72b23-5, see 47,21ff. (where we find a more accurate quote).

Interestingly, in our passage Philoponus gives an ontological reading of the term *horoi* much closer to Ammonius' interpretation, which he there rejected in favour of Themistius' more logically oriented reading. See the next note for a difference between Ammonius and Philoponus.

502. The point here is that the intellect can be called the beginning of *epistêmê* even if it is not itself *epistêmê*, because insofar as it gives us the *principles* of scientific understanding, its 'lowest' limit touches the lower form of cognition. Whereas Ammonius, in Philoponus' paraphrase, identified *nous* as the divine intellect (47,25-6), Philoponus himself clearly takes it to be the human intellect as a capacity of the soul, hence 'that (*toutou*) intellect'. This also explains that he can speak of an activity of intellect: the human intellect is not always actual. 'Lowly' translates *peripezios*, an uncommon word used by late Neoplatonic commentators to indicate that which is closer to matter (as in Philop. *in Phys.* 220,17 about the lowest part of mathematics being related to science of nature; Simpl. *in Ench.* 126,3-6 on the doxastic part of the soul which is concerned with matter).

503. Since Philoponus is here discussing *doxa* and *epistêmê* not as dispositions but as capacities of the soul (cf. 323,28-31), he can speak of their activities.

504. Considering that Philoponus should still be speaking of the capacity of *doxa*, the expression he uses here, *tini doxei*, is somewhat surprising. We have decided to assume that Philoponus did not change the subject to the result of the activity of the capacity (i.e. *doxa* in the sense of 'an opinion'), and leave *tini* untranslated.

505. For consistency we are holding onto 'scientific understanding' here, despite the fact that from this point onwards Aristotle uses *epistêmê* in a less narrow and technical sense.

506. i.e. perishing.

507. Ross, unlike the lemma, does not include *ouk*, rendering the translation 'So in what sense is having an opinion the same ...'.

508. 'Believe' translates *doxazein*, which we more commonly translate as 'have an opinion'.

509. This paragraph seems completely superfluous.

510. As Mignucci, 656, points out, this use of the term *akolouthêsei* can be found in Plato, *Phd.* 107B6-7.

511. Ross punctuates 89a22-3 as a question.

512. Philoponus' suggestion to supply 'of the reason why' is based on a misunderstanding of the phrase 'both of the fact and of the reason why'. He takes that phrase to go with the foregoing 'he will have an opinion and he will not truly have scientific understanding', where, however, it is irrelevant. Instead, it goes with the subsequent 'if he has an opinion through immediates': on the one hand we have opinion through immediates, which is both of the fact and the reason why, and on the other hand we have opinion which is not through immediates and which is therefore only of the fact.

513. This is probably a reference to 89a2-3.

514. Reading the full stop before *kath'ho* as a colon.

515. i.e. that [the soul] is *not* immortal.

516. There's a lacuna in the text, and Wallies proposes *tauton toi mê doxazein athanaton einai*, 'would be the same as not being of the opinion that it is immortal'. This is indeed what we want to end up with, in order for the example to be an example of someone 'not having the opinion that he has'. However, in that case there is a missing link. 'Thinking that the soul is immortal' is true, and if the truth (i.e. that which is true) is identical to the false, then 'thinking that the soul is immortal' is identical to 'thinking that the soul is *not* immortal'. We thus obtain an example of the first problem, namely that the true opinion would be the same as the false one. Only if we subsequently take the latter to be equivalent to 'not being of the opinion that the soul is immortal', do we also have an example of 'not having the opinion that one has'. Moreover, Philoponus misunderstands Aristotle, for whom the paradox is that of not having an opinion of that of which we have a *false* opinion (in the example: that the soul is immortal). As Barnes, 191, points out, it is not entirely clear what Aristotle means here. He suggests that we read it as follows: if we have a false opinion about the soul being immortal, we do not really have an opinion about it, because it is in fact true that the soul is immortal.

517. Accepting Wallies' insertion of *to*.

518. Accepting Wallies' insertion of *asummetros hê diametros kai tên legousan hoti*.

519. Translating according to Philoponus' understanding of the text (see 330,5), although a more proper translation would probably be 'it is the same, because human being [is the same]'.

520. It is interesting that Philoponus does not mention studying the celestial phenomenon, but talking to people who did. This suggests a view of scientific understanding that is close to that of Barnes (1969), namely as (primarily) pedagogical and transferable from one person to another through talking.

521. Prudence (*phronêsis*) and craft (*tekhnê*): esp. *EN* 6.5-13.

522. Note that Philoponus takes *Alpha Elatton* not to be part of the *Metaphysics*. For a brief overview of modern views of the status of *Alpha Elatton*, and for Alexander's position, see Dooley and Madigan (1992), 3-4. At 141,8 Philoponus refers to *Metaph. Gamma* as Book 3 of the *Metaphysics*.

523. Intellect (*nous*) and wisdom (*sophia*): Wallies refers to *Metaph.* 1.1-2 and *alpha elatton* 993b9ff. One might add 994b14-16. Scientific understanding (*epistêmê*), outside the *Posterior Analytics*, thought (*dianoia*) and opinion (*doxa*): *DA* 3.3. Interestingly, Philoponus does not refer to the most important Aristotelian discussion of intellect, in *DA* 3.4 and 5. Apparently, it is more important for Philoponus to have the discussion of intellect and wisdom be a part of theology, than to be accurate.

524. The conjunction in question is the apparent conjunction of moon and sun, i.e. new moon. The only other reference we have found to the moon having blunt horns and impending rain is *Scholia in Aratum* 785,16-18.

525. Although grammatically *hai praxeis* can fall under *pantes*, 'craftsmen and actions' is a surprising combination. Considering the foregoing, we can take 'actions' to stand for that which uses prudence (*phronêsis*), just as the craftsmen use craft.

526. Assuming that the accusatives are in anticipation of *dihelein bouloito* in line 6.

527. i.e. at *An. Post.* 89b8.

528. This piece of Neoplatonic etymologising apparently gains a tiny bit of popularity in the sixth century. It occurs also in Asclepius' commentary on Nicomachus (1,1,7-8) and in Pseudo-David/Pseudo-Elias on the *Isagoge* (17 30,22). Note that earlier, at 324,5ff., Philoponus distinguished between intellect itself, thanks to which we know the divine (i.e. the function here ascribed to *sophia*), and 'a last and lowly activity of (our) intellect' thanks to which we grasp common notions and immediate premises (here ascribed to intellect itself).

529. Philoponus is referring to *Tim.* 52B, where Plato introduces 'bastard reasoning' as the capacity which knows matter. On matter without form being unknowable cf. *Metaph.* 7.10 1036a8-9 and on it being formless, Porph. *Sent.* 20.

530. This is a combination of different themes (*Metaph.* 2, 993b9-11 on the day-blindness of bats, *DC* 290a13-18 on the sun vs. other stars and the weakness of human sight, and on human weakness and knowledge of the divine), into one *topos*, also found in Ascl. *in Metaph.* 114,31ff., 117,26ff.; David, *Prol.* 46,20-5; Olymp., *in Gorg.* 30 3,19-22; Elias, *in Isag.* 24,5-7).

531. *Klêroun* is very rarely construed with a genitive, but *toutou tou onomatos keklêrôtai* may perhaps be explained as an abbreviation of *to tês sophias onoma keklêrôtai.*

532. i.e. *An. Post.* 1.3, 72b4 and 2.19, 100b15. Cf. above 324,6ff. (where in light of the context we translated *horoi* as 'definitions') and Philoponus' comments on *An. Post.* 1.3 at 47,21ff.

533. The categories are principles of scientific understanding/syllogism in the sense that without the former, the latter is impossible. Cf. Philop. *in Cat.* 10,24ff. Philoponus' ascription of the discovery of the categories to intellect is echoed in Sophonias' paraphase of *in DA* (122,38ff.), which suggests that it was in Philoponus' commentary on *DA* 3.

534. 'In an imperceptible time' translates *askeptos*, a term which usually has a pejorative sense. Cf. Mignucci, 673.

535. Adding commas before and after *phêsin*, to prevent the conditional subclause from having two unconnected main verbs (*phêsin* and *apokrineitai*).

536. Note that Philoponus calls acumen an *energeia*, rather than a *dunamis* (as opposed to Them., *in An. Post.* 41,5 and Aristotle himself, *Rhet.* 1362b24-5).

537. It may seem that Philoponus gets the example backward, but that is not the case: since acumen is about spotting the middle term, Aristotle must mean that the acute person sees a poor man talk to a rich man and spots the reason, or more precisely, the final cause (cf. Detel II 541): he does this in order to borrow money. The relation between middle term and final cause is problematic, but the ancients seem unaware of this fact. On this issue see Leunissen (2007).

538. The element of being initiated together (*summustai*) is an interesting addition. The term is late ancient, has its source in the mysteries (see e.g. the so-called 'Mithras Liturgy' in *PGM* IV, l.732 (*ANRW* vol. I, 3497), and is subsequently adopted by Christians to refer to those who are baptised (see Inge [1899], app. 2).

539. We do not know which Platonic passage Philoponus has in mind. The thought that friendship transcends justice because the former does not need the latter whereas the latter does need the former is found at *EN* 1154b22-7. Friendship as unifying and divine possession is found also at Simpl. *in Epict. Ench.* 89,8ff.

540. Note that Philoponus' little digression on friendship disqualifies Aristotle's example: apparently the acute person in the example identifies as a cause of friendship something (namely sharing an enemy) which is not a cause of true friendship.

541. Ross brackets the article at 89b15, referring to P, who leaves it out at 333,26.

542. 'Outline' translates *diagramma*, which the commentators use on occasion to refer to a schematic rendering of syllogisms. Cf. e.g. Philop., *in An. Pr.* 178,19 and Alex., in *An. Pr.* 406,12.

Bibliography

Anton, J. and Preus, A. (eds) (1990), *Essays in Ancient Greek Philosophy*, V: *Aristotle's Ontology*, Albany.

Barker, A. (2007), *The Science of Harmonics in Classical Greece*, Cambridge.

Barnes, J. (1969), 'Aristotle's Theory of Demonstration', *Phronesis* 14(2), 123-52.

Barnes, J. (1981), 'Proof and the Syllogism', in E. Berti (ed.), 17-59.

Barnes, J. (1994), *Aristotle: Posterior Analytics*, Clarendon Aristotle Series, 2nd edn, Oxford.

Berti, E. (ed.) (1981), *Aristotle on Science: The 'Posterior Analytics'*, Padua.

Burnyeat, M.F., 'Aristotle on Understanding Knowledge', in E. Berti (1981), 97-139.

De Haas, F.A.J. (1999), Review of D.A. Di Liscia, E. Kessler and C. Methuen (eds) (1997), in *Renaissance Studies* 13(3), 349-52.

Detel, W. (1993), *Aristoteles Analytica Posteriora*, Berlin.

Di Liscia, D.A., Kessler, E. and Methuen, C. (eds) (1997), *Method and Order in Renaissance Philosophy of Nature: The Aristotle Commentary Tradition*, Aldershot-Brookfield.

Dooley, E. and Madigan, A. (1992), *Alexander of Aphrodisias: On Aristotle Metaphysics 2-3*, London.

Frede, M. (1987), *Essays in Ancient Philosophy*, Oxford.

Frede, M. 'Stoic vs. Aristotelian Syllogism', in M. Frede (1987), 99-124, reprint of *Archiv für Geschichte der Philosophie* 56 (1974), 1-32.

Goldin, O. (1996), *Explaining an Eclipse: Aristotle's Posterior Analytics 2.1-10*, Ann Arbor.

Goldin, O. (1999), *Philoponus(?): On Aristotle Posterior Analytics 2*, London.

Goldin, O. (2010), 'Two Traditions in the Ancient *Posterior Analytics* Commentaries', in F.A.J. de Haas, M. Leunissen and M. Martijn (eds), 155-82.

Haas, F.A.J. de, Leunissen, M. and Martijn, M. (eds) (2010), *Interpreting Aristotle's Posterior Analytics in Late Antiquity and Beyond*, Leiden.

Heath, T.L. (1956), *Euclid. The Thirteen Books of the Elements*, New York.

Inge, W.R. (1899), *Christian Mysticism*, London.

Jaeger, W. (1960), 'Emendationum Aristotelearum specimen', in *Scripta minora*, Rome.

Leunissen, M.E.M.P.J. (2007), 'The Structure of Teleological Explanations in Aristotle: Theory and Practice', *Oxford Studies in Ancient Philosophy* 33, 145-78.

Lloyd, A.C. (1990), *The Anatomy of Neoplatonism*, Oxford.

Martijn, M. (2010), *Proclus on Nature: Philosophy of Nature and Its Methods in Proclus' Commentary on Plato's Timaeus*, Leiden.

McKirahan, R. (1992), *Principles and Proofs, Aristotle's Theory of Demonstrative Science*, Princeton.

McKirahan, R. (2008), *Philoponus: On Aristotle Posterior Analytics 1.1-8*, London.

McKirahan, R. (2009), 'Philoponus' Account of Axioms in his Commentary on

Aristotle's *Posterior Analytics*', *Documenti e Studi Sulla Tradizione Filosofica Medievale* 20, 211-63.
McKirahan, R. (2012), *Philoponus: On Aristotle Posterior Analytics 1.9-18*, London.
Mignucci, M. (1975), *L'argomentazione dimostrativa in Aristotele. Commento agli Analitici Secondi*, vol. 1, Padua.
Morrison, D. (1997), 'Philoponus and Simplicius on Tekmeriodic Proof', in D.A. Di Liscia, E. Kessler and C. Methuen (eds), 1-22.
O'Meara, D.J. (1989), *Pythagoras Revived*, Oxford.
Perelmuter, Z. (2010), '*Nous* and Two Kinds of *Epistêmê* in Aristotle's *Posterior Analytics*', *Phronesis* 55, 228-54.
Richard, M. (1950), '*Apo phones*', *Byzantion* 20, 191-222.
Richardson Lear, G. (2004), *Happy Lives and the Highest Good*, Cambridge.
Ross, W.D. (1949), *Aristotle's Prior and Posterior Analytics*, Oxford.
Schofield, M. and Nussbaum, M. (eds) (1982), *Language and Logos*, Cambridge.
Share, M. (2005), *Philoponus: Against Proclus On the Eternity of the World 1-5*, London.
Solsen, F. (1929), *Die Entwicklung der aristotelischen Logik und Rhetorik*, Berlin.
Sorabji, R. (1982), 'Myths about Non-Propositional Thought', in M. Schofield and M. Nussbaum (eds), 295-314.
Sorabji, R. (ed.) (1990), *Aristotle Transformed: The Ancient Commentators and Their Influence*, London.
Sorabji, R. (2005), *The Philosophy of the Commentators 200-600 A.D.: A Sourcebook* (3 vols), London.
Sorabji, R. (2006), 'Universals Transformed: The First Thousand Years After Plato', in P.F. Strawson and A. Chakrabarti (eds), 105-26.
Sorabji, R. (2010), *Philoponus and the Rejection of Aristotelian Science*, London.
Strawson, P.F. and Chakrabarti, A. (eds) (2006), *Universals, Concepts and Qualities: New Essays on the Meaning of Predicates*, Burlington.
Temporini, H. and Haase, W. (eds) (1973-), *Aufstieg und Niedergang der römischen Welt*, Berlin.
Urbanas, A. (1990), 'On "Essentially" (*hoper*) in Aristotle', in J. Anton and A. Preus (eds), 95-110.
van der Eijk, P. (2005), *Philoponus: On Aristotle On the Soul 1.1-2*, London.
Verrycken, K. (1990a), 'The Metaphysics of Ammonius son of Hermeias', in R. Sorabji (ed.), 199-231.
Verrycken, K. (1990b), 'The Development of Philoponus' Thought and its Chronology', in R. Sorabji (ed.), 233-74.
Verrycken, K. (1994), *De vroegere Philoponus. Een studie van het Alexandrijnse Neoplatonisme*, Brussels.
Verrycken, K. (1998), *Alexandrië 529 – Philoponus en het einde van de antieke filosofie*, Budel.
Wallies, M. (1909), *Ioannis Philoponi in Aristotelis Analytica Posteriora Commentaria*, *CAG* 13.3, Berlin.

English-Greek Glossary

a certain kind: *poion*
a finite number of times: *peperasmenakis*
a fortiori: *ek pollou tou periontos, pollôi pleon*
a number of things: *pleiôn*
above: *anô*
absolute: *haplôs*
absurd: *atopos*
accept in addition: *epidekhesthai*
accident: *sumbainon, sumbebêkos*
accidentally: *kata sumbebêkos*
according to: *hoson epi*
account: *apodosis*
accusation: *aitia*
action: *poiein, praxis*
activity: *energeia*
actuality: *energeia*
acumen: *ankhinoia*
acute: *ankhinous*
add: *prostithesthai*
add by apposition: *proslambanein*
addition: *prosthêkê, prosthesis*
adjacent: *ephexês, prosekhês, sunekhês*
adjacently: *prosekhôs*
adjoin: *episunaptein*
admit: *sunkhôrein*
advance: *proodos*
affirm: *kataphaskein*
affirmation: *kataphasis*
affirmative: *kataphatikos, katêgorikos*
affording demonstration: *apodeiktikos*
agora: *agora*
agree: *homologein, sunkhôrein*
aim for: *apidein*
aim: *skopos*
aimed at: *skopimos*
akin, i.e. of the same genus: *sungenês*
all: *pas*
already subsist: *prohuparkhein*
alter: *metapiptein*
always in motion: *aeikinêtos*
analyse: *analuein*
analysis: *analusis*
analytic: *analutikos*
ancestor: *progonos*
ancient: *arkhaios*
and so forth: *ephexês*
angle: *epibolê*
angle: *gônia*
animal: *zôion*
antecedent: *hêgoumenon*
any: *pas*
apart from: *khôris*
apparent: *phaneros, phanos*
appearances: *ta phainomena*
apply: *epharmottein, epharmozein, epipherein, harmozein, metapherein*
apprehend: *antilambanein, epiballein*
apprehension: *antilêpsis, hupolêpsis, katalêpsis*
approach: *prosienai*
appropriate: *oikeios, prosphuês*
appropriately: *prosêkontôs*
approximate: *engus*
area: *topos*
argue against: *elenkhein*
argument: *epikheirêma, logos*
argument that establishes: *kataskeuastikos*
arise: *gignesthai*
arithmetic: *arithmêtikê*
arithmetician: *arithmêtikos*
arrange: *tattein*
arrive at: *katantan, phthasai* (+ *eis*)
arrive at truth: *alêtheuein*
article: *arthron*
articulate: *diarthroun*
as an additional result: (*ek*) *periousias*
as is his wont: *sunêthôs*
as many as: *isarithmos*
as such: *haplôs*
ascend: *anerkhesthai, anienai*
ascribe: *anatithenai*
ask: *erôtan*
assert: *legein*
assign: *apodidonai, aponemein*
assign names: *onomatothetein*
assume: *hupolambanein, lambanein*
assume in addition: *proslambanein*
astronomy: *astronomia*

at all: *haplôs, holôs*
at all events: *pantôs*
at any rate: *pantôs*
at hand: *prokeimenos*
at once: *euthus*
at the same rate: *isarithmôs*
at the same time: *hama*
attack: *epilambanein, epirrapizein*
attend to: *prosballein*
attribute (n.): *pathos, sumptôma*
attribute (v.): *anagein*

badly put: *akatallêlos*
bald: *phalakros*
base: *basis*
bastard: *nothos*
bat: *nukteris*
bathe: *louesthai*
be: *einai, gignesthai, hupokeisthai*
be a fact: *tunkhanein*
be accustomed: *êthein*
be acted on in turn: *antipaskhein*
be active: *energein, dran*
be added: *proskeisthai*
be alive: *zôein*
be altered: *alloiousthai*
be an accident of: *sumbainein*
be an apodosis: *antapodidonai*
be associated: *koinonein*
be bounded: *peraiousthai, perainein*
be close to: *engizein*
be coextensive: *antistrephein, exisazesthai*
be composed: *sunkeisthai, sunistanai*
be concerned with: *prokeisthai* (+ dat.)
be conjoined with: *epipheresthai*
be consistent with: *sunhepesthai*
be convinced: *hupolambanein, pistousthai*
be dark: *skiazein*
be deflected: *kelasthai*
be distinct: *diôrizesthai*
be distinguished from one another: *antidiaireisthai*
be engaged in dialogue: *prosdialegesthai*
be equivalent to: *isodunamein*
be evidenced: *pistousthai*
be false: *pseudesthai*
be finite: *perainein*
be generated: *gignesthai*
be given: *tunkhanein*
be happy: *eudaimonein*
be healthy: *hugiainein*
be in dispute: *amphiballein*
be in motion: *kineisthai*
be in opposition: *enantiotithenai*
be inductive: *epagein*
be lacking: *ekleipein, epileipein*
be left over: *perileipesthai*
be mistaken: *apatêsthai*
be mixed with: *epimignunai*
be mutually predicated: *antikatêgoreisthai*
be next to: *ekheisthai*
be obvious: *phainesthai*
be opposed: *antidiaireisthai, antikeisthai*
be posited: *hupokeisthai*
be possible: *endekhesthai, enkhôrein*
be preferred: *hairetos*
be present: *keisthai, pareinai*
be put forward: *prokeisthai*
be puzzled: *aporein*
be realised: *ekbainein*
be reversed: *antistrephein*
be said to occur: *legein*
be satisfied: *agapein*
be situated: *keisthai*
be suitable: *prepein*
be the case: *huparkhein*
be the interlocutor: *prosdialegesthai*
be true: *alêtheuein*
be true simultaneously: *sunalêtheuein*
be unaware: *agnoiein*
be up to: *poiein*
be useful: *khrêsimeuein*
be varied: *poikillesthai*
be visible: *phainein*
bear on: *sunteinein pros, teinein pros*
beast of burden: *hupozugios*
beautiful: *kalos*
become or be acquainted with: *gnôrizein*
bed: *klinê*
before: *proteros*
before one: *prokeimenos* (+ dat.)
begin by: *prôton …*
begin by assuming: *prolambanein*
begin by specifying: *prodiorizesthai*
beginning: *arkhê*
behold: *theasthai*
being a more important cause: *aitiôteros*
being a more important explanation: *aitiôteros*
being deceived: *apatê*
being evident: *enargeia*
believe: *doxazein, hêgeisthai, pisteuein*
belly: *gastêr*
belong: *huparkhein*
belonging to the art: *tekhnikos*

below: *katô*
bend: *neuein*
better: *beltiôn*
between: *metaxu*
beyond: *peraiteros*
black: *melas*
block: *antiphrattein*
blunt: *amblus*
boat: *naus*
bodiless: *asômatos*
body: *sôma*
borrow: *dianeizesthai*
both hold: *sunhuparkhein*
boundless: *aperilêptos*
bring along: *epipheresthai*
bring forward: *epagein*
bring to rest: *êremizein*
bronze: *khalkos*
burn: *kaiein*
buy: *ôneisthai*
by all means: *pantôs*
by direct apprehension: *epiblêtikôs*
by itself: *idiai*
by nature: *kata phusin, pephukos* (+ inf.)
by parallel reasoning: *ek parallêlou*
by-product: *parakolouthêma*

call: *kalein, legein, phanai*
camp: *stratopedon*
can: *endekhesthai*
can be otherwise: *endekhesthai*
capable of apprehending: *antilêptikos*
capable of laughter: *gelastikos*
capacity: *dunamis*
capacity of hitting upon: *eustokhia*
carat: *keration*
carpentry: *tektonikê*
case: *ptôsis*
categorical: *katêgorikos*
category: *katêgoria*
causal: *aitiologikos*
cause: *aitia, aition*
caused: *aitiatos*
cease: *istasthai*
centre: *kentros*
chance: *tukhê*
change: *ameibein, metaballein*
change along with: *summetaballein*
change one's mind: *metadoxazein*
circle: *kuklos*
circular: *kuklôi*
clarify: *dêloun, saphênizein*
clarity: *saphêneia, saphia*
clash with: *antipiptein*
class: *meros*
classify: *anagein*
clear: *phaneros, saphês*
clearly: *enargôs*
close: *engus*
cold: *psukhros*
collect: *sunagein*
colour: *khrôma*
combination: *sumplokê, sunthesis*
combine: *sumplekein*
combine successively: *episuntithenai*
come around: *periagein, perierkhesthai*
come to: *katantan, paragignesthai*
come to a stop: *histasthai*
come to light: *anaphainein*
come upon: *katantan*
comma: *hupostigmê*
commensurability: *summetria*
commensurable: *summetros*
commentator: *exêgêtês*
commit an error: *hamartanein*
common: *koinos*
compare: *sunkrinein*
complete (adj.): *entelês*
complete (v.): *plêroun, suntelein*
completely: *holôs*
composite: *sunthetos*
compound: *sumpeplegmenos*
compounded: *miktos*
comprehend: *periekhein*
concede: *sunkhôrein*
conceive: *kuein, kuophorein*
conceive of: *noein, epinoein*
concern: *katagignesthai peri*
conclude: *sumperainesthai, sunagein*
conclusion: *sumperasma*
concord: *sumphônia*
concrete: *pragmateiôdês*
confirm: *marturein*
conjunction: *sundesmos, sunodos, suzugia*
consequent: *hepomenon*
consider: *blepein, episkeptesthai, theôrein*
consider in addition: *proslogizesthai*
consist of: *sunkeisthai*
construct: *kathistanai*
construction: *suntaxis*
contingently: *endekhomenôs*
continuous: *sunekhês*
continuously: *kata sunekheian*
continuum: *sunekheia, sunekhes*
contradiction: *antiphasis*
contrast (n.): *antidiastolê*
contrast (v.): *antitithenai*

contribute: *sumballein*
converse: *dialegesthai*
conversion: *antistrophê*
convert: *antistrephein*
convincing: *pithanos*
correlatively: *hama*
counter-objection: *antiparastasis*
countless: *murios*
cowardice: *deilia*
craft: *tekhnê*
craftsman: *tekhnitês*
credibility: *pistis, to piston*
cross: *metabainein*
crow: *korax*
cubit: *pêkhus*
cultured: *mousikos*
curved: *kampulos*
custom: *sunêtheia*
cut: *temnein*

dative: *dotikos*
deal with: *khrêsthai, lambanein*
death: *thanatos*
decide: *bouleusthai*
declare: *apophainesthai*
deduce: *sullogizesthai*
deduction: *sullogismos*
deem: *axioun*
defend oneself: *apologeisthai*
defence: *apologia*
define: *horizesthai*
definition: *horos, logos*
definitional: *horistikos*
delimit: *perigraphein*
demiurgic: *demiourgos*
demonstrable: *apodeiktos*
demonstration: *apodeixis*
demonstrative: *apodeiktikos*
deny: *apophanai, apophaskein, sterein*
depart: *aperkhesthai*
derive credibility from: *pistos einai ek*
derive its conviction in … from …: *pistousthai* (+ acc. & *ek*)
describe: *erein, kalein*
description: *hupographê*
determinate: *hôrismenos*
determinately know: *diaginôskein*
determination: *prosdiorismos*
determine: *aphorizein*
develop: *poiein*
develop an argument: *logon poieisthai*
develop the argument: *kataskeuazein*
development of the argument: *kataskeuê*
devoid: *kenos*
dialectical: *dialektikos*
dialectically: *dialektikôs*
diagonal: *diametros*
differ: *diapherein, diistanai*
difference: *diaphora*
different: *diaphoros*
differentia: *diaphora*
direct refutations at: *enistanai*
direct(ly): *ep' euthuias*
discern: *antilambanein*
discover: *heuriskein*
discovery: *heuresis*
discrete: *diôrismenos*
discuss: *dialegesthai*
discussion: *logos*
dispersive: *diakritikos*
display: *endeiknusthai*
distinction: *diaphora*
distinguish: *diairein, diakrinein, diôrizein*
distribute: *dianemein*
divide: *diairein, diôrizein*
divine: *theios*
divisible: *diairetos*
division: *diairesis*
do: *poiein*
do battle: *makhesthai*
doctrine: *dogma, doxa*
dog: *kuôn*
downwards: *anôthen, katô*
draw a conclusion: *sumperainesthai, sunagein, sumperasma sunagein*
draw out: *katateinein*
dry area: *xêra*

earth: *gê*
easy to distinguish: *diadêlos*
eat: *phagein*
ebony: *ebenos*
eclipse (n.): *ekleipsis*
eclipse (v.): *ekleipein*
eclipse (v., intrans.): *eklimpanein*
efficient: *poiêtikos*
eighteen times: *oktôkaidekaplasion*
eighth: *ogdoos*
elaborate separately on: *dialambanein* (+ *peri*)
element: *stoikheion*
eliminate: *anairein*
eliminate together with: *sunanairein*
elimination: *anairesis*
emit: *proienai*
employ: *katakhrasthai, khrêsthai, lambanein, paralambanein*

encompass: *periekhein*
encompassing: *periektikos*
encounter: *entunkhanein*
end: *eskhatos, telos*
end in: *katalêgein*
enemy: *ekhthros*
enter: *emballein*
enterprise: *pragmateia*
entirely: *pampan, pantôs*
entirely necessary: *pas anagkê*
enumerate: *aparithmein, katarithmein*
equal: *isos*
equality: *isotês*
equivocal: *homônumos*
equivocally: *kath' homônumian*
equivocity: *homônumia*
error: *hamartêma*
essence: *ousia, to ti ên einai*
essential: *kat' ousian, kata tês ousias, ousiôdês*
essentially: *kat' ousian, ousiôdôs*
establish: *enkataskeuazein, kataskeuazein*
eternal: *aidios*
ethics: *êthikê*
even: *artios*
ever: *pote*
every: *pas*
evidence: *sêmeion*
evident: *enargês, phaneros*
example: *hupodeigma, paradeigma*
excluding: *diakrisis*
exist: *einai*
explain: *exêgeisthai*
explanation: *aitia*
explicitly: *diarrêdên*
exposition: *ekthesis*
expound: *ektithenai*
extend: *ekballein, epekteinein*
extend further: *huperteinein, pleon ekhein*
exterior: *ektos*
extreme: *akros*

fabricate: *dêmiourgein*
fact: *pragma, to hoti*
fact of physics: *phusikon*
fall: *piptein*
fall outside: *eksô piptein*
fallacy: *paralogismos*
false: *pseudos*
falsehood: *pseudos*
falsely called: *pseudônumôs*
fault: *hamartêma*
feel pleasure: *hêdesthai*
female: *thêlus*
figure: *skhêma*
final: *telikos*
find: *heuriskein*
finite: *hôrismenos*
finite in number: *peperasmenos*
fire: *pur*
first: *proteron, prôtos, prôtistos*
first ...: *phthasai* (+ part.)
first determine: *prosdiorizesthai*
first word: *arkhê*
five: *pente*
flatnailed: *platuonukhos*
follow: *akolouthein, leipesthai, sumbainein*
foot: *pous*
for certain: *bebaiôs*
for example: *ei tukhoi*
for now: *teôs*
for the sake of completion: (*ek*) *periousias*
form: *eidos*
form an essential part of: *sumplêrein, sumplêrôtikos*
formal: *eidikos, logikos*
formally: *logikôs*
formless: *aneideos*
form-producing: *eidopoios*
four-sided figure: *tetrapleuron*
friend: *philos*
friendship: *philia*
from above: *anôthen*
from below: *katôthen*
from outside: *exôthen*
fulfil: *apotelein*
fully work through: *epexerkhesthai*
further: *pleiôn*

general: *koinos*
generally agree upon: *sungignôskein*
generally speaking: *holôs*
generate: *gennan*
generic: *genikos*
genus: *eidos, genos*
geometer: *geômetrês*
geometrically: *geômetrikôs*
geometry: *geômetria*
get confused: *planasthai*
give: *epitithenai, tithenai*
give an example: *paradeiknusthai*
give credibility: *pistousthai*
give precise information: *akribologein*
give the place of: *tattein*
given: *lephthen*

glass: *huelos*
go forward: *probainein, proerkhesthai*
go on: *proerkhesthai*
go on foot: *pezeuein*
go on to say: *epagein*
go out: *proerkhesthai*
go through: *dierkhesthai, diexerkhesthai, epexienai*
go wrong: *hamartanein*
goatstag: *tragelaphos*
god: *theos*
good: *khrêma*
grammarian: *grammatikos*
grammatical: *grammatikos*
grant: *didonai, sunkhôrein*
grasp (v.): *ekhein, lambanein, paralambanein, hairein*
grasp (n.): *lêpsis*
grasp with intellection: *noein*
grave: *taphos*
guideline: *kanon*

hand: *kheir*
happen: *sumbainein*
happen to be: *tunkhanein*
happen to: *paskhein*
hard: *sklêros*
harmonics: *harmonikê*
have: *ekhein*
have a position: *hupokeisthai, keisthai*
have allotted: *klêrousthai*
have already ...: *phthasai* (part. + aor.)
have an opinion: *doxazein*
have scientific understanding: *epistasthai*
have things in common: *koinônein*
having a spherical shape: *sphairoeidês*
having scientific understanding: *epistêmôn*
he devotes some time: *teôs*
he who has scientific understanding: *epistêmôn*
hear: *akouein*
hearing: *akoê*
heaven: *ouranos*
heavenly: *ouranios*
heavy: *barus*
high: *akros, anô, oxus*
hold: *huparkhein*
hold ground: *khôran ekhein*
honour: *timân*
horn: *keras*
horse: *hippos*
house: *oikia*
how much: *poson*
how we should construct: *katastateon*
human being: *anthrôpos*
humanity: *anthrôpotês*
hypothesis: *hupothesis*
hypothesise: *hupotithenai*
hypothetical: *hupothetikos*

Idea: *idea*
ignorance: *agnoia*
illuminate: *phôtizein*
illustrate: *phôtizein*
immaterial: *aulos*
immediate: *amesos*
immediately: *autothen, ephexês*
immortal: *athanatos*
imperceptible: *askeptos*
imperishability: *aphtharsia*
imperishable: *aphthartos*
impossible: *adunatos*
in a circle: *kuklôi*
in a common manner: *koinôs*
in a continuous manner: *kata sunekheian*
in a summary fashion: *suntomôs*
in a way that identifies the explanation: *aitiologikôs*
in act: *energeiai*
in actuality: *energeiai, kat' energeian*
in an unnatural manner: *para phusin*
in between: *metaxu*
in common: *koinôs*
in detail: *platu*
in every instance: *pantôs*
in every sense: *pantôs*
in general: *haplôs, holôs, katholou, koinôs*
in good order: *errômenos*
in how many ways: *posakhôs*
in inverted order: *huperbatôs*
in itself: *kath' heauto*
in more than one way: *pleonakhôs*
in most cases: *hôs epi to polu*
in one dimension: *eph' hen diastaton*
in precisely the same manner: *aparallaktôs*
in question: *prokeimenos*
in reality: *tôi onti*
in some way or another: *toiôs ê toiôs*
in the general sense: *haplôs*
in the opposite way: *anapalin*
in the other direction: *anapalin*
in the same series of predication: *sustoikhos*
in the strict sense: *haplôs, holôs, kuriôs*

inarticulate: *anarthros*
inasmuch as: *hoson epi*
include: *paralambanein*
incommensurable: *asummetros*
incorrectly: *kakôs*
increase: *auxanein*
increase together with: *sunauxanein*
indeed: *pantôs*
indefinite: *aoristos, apeiros*
indemonstrable: *anapodeiktos*
independent: *autarkês*
independently: *idiai*
indeterminate: *aoristos*
indicate: *sêmainein*
indifferent: *adiaphoros*
individual: *atomos, kath' hekastos, monadikos*
individually: *idiôs*
induce: *sunagein*
induction: *epagôgê*
infer: *sunagein*
inferior: *kheirôn*
infinite in number: *apeiros*
infinitude: *apeiria*
infinity: *apeiron*
inhere: *enuparkhein*
initiate: *kinein*
injustice: *adikia*
inquiry: *skepsis, zêtêsis*
insert: *emballein, paremballein, parentitesthai*
insert a comma: *hupostizein*
inside: *entos*
instantly: *en akariaiôi*
instil: *empoiein*
intellect: *nous*
intellection: *noêsis*
intellectual: *noeros*
intelligible: *noêtos*
intemperance: *akolasia*
intend: *boulesthai*
interior: *entos*
intermediate: *metaxu, mesos*
interposition: *antiphraxis*
interpret: *akouein*
interpretation: *exêgêsis*
intertwine: *sumplekein*
interval: *diastêma*
introduce: *epagein*
introductory chapter(s): *prooimion*
intuition: *epibolê*
investigate: *episkeptesthai, zêtein*
invisible: *aoratos*
involve: *ekhein*
irrational: *alogos*
irrelevant to: *ouden pros*
is distant from: *aphistasthai*
is to be sought: *skepteon*
isolate: *dialambanein*
isosceles: *isoskelês*
it is no surprise that: *eikotôs*
it is reasonable that: *eikotôs*
it is required: *opheilon einai*
it is to be expected that: *eikotôs*
it must be pointed out: *sêmeiôteon*
it so happens that: *houtôs tukhon*
itself by itself: *auto kath' hauto*

jointly: *koinôs*
just: *dikaios*
justice: *dikaiosunê*

keep: *phulattein*
kind: *eidos, genos*
kinship: *sungeneia*
kithara: *kithara*
kithara player: *kitharôidos*
know: *eidenai, ginôskein*
know in advance: *proginôskein*
knowable: *gnôstos*
knowledge: *gnôsis*
known: *gnôrimos*

lack (n.): *ekleipsis*
lack (v.): *leipesthai*
lantern: *lamptêr*
last: *eskhatos, hustatos*
later: *husteros*
lay down a rule: *nomotithenai*
lead to: *sunagein*
lead to a conclusion: *sumperasma, sunagein*
lead up: *anagein*
learn: *manthanein*
leave off at: *katalêgein, lêgein*
letter: *stoikheion*
lifeless: *apsukhos*
light: *phôs*
like: *toioutos*
limit: *peras*
limited: *peperasmenos*
line: *grammê*
link: *episunaptein*
living: *empsukhos*
logical: *logikos*
logos: *logos*
long knife: *makhaira*
look into: *skopein*

low: *barus*
lower: *hupokatô*
lowly: *peripezios*
luminosity: *to lampron*

magnetic: *magnêtis*
magnitude: *megethos*
major: *meizon*
make: *ergazesthai, poiein*
make an additional point: *prostithesthai*
make an objection: *enistasthai*
make distinctions: *diarthroun*
make one's way: *badizein*
make perfect: *teleioun*
make sense: *ekhein tina logon*
make use of: *khrêsthai, khrêzein*
male: *arren*
man: *anthrôpos*
manner: *tropos*
manuscript: *antigraphon*
many times: *pollakis*
material: *hulikês*
mathematics: *mathêmata*
matter: *hulê*
mean: *erein, episêmainein, phanai, sêmainein*
mean when saying that: *houtôs phanai*
mediated: *emmesos*
medical: *iatrikos*
meet: *haptesthai*
mention: *eipein, erein, katalegain, mimnêskesthai*
metal: *metallon*
method: *ephodos*
method of proof: *agôgê*
methodically: *diexodikôs*
middle: *mesos*
middle term: *mesos*
minor: *elatton*
minor assumption: *lêmmation*
miscellaneous: *summiktos*
mistake: *apatê*
misunderstand: *eklambanein kakôs*
mna (unit of weight): *mna*
moderation: *sophrosunê*
mold: *diaplassein*
moon: *selênê*
more: *pleiôn*
more than one: *pleiôn*
mortal: *brotos, thnêtos*
most people: *hoi polloi*
mother: *mêtêr*
motion: *kinêsis*
move: *kineisthai*
move on: *metabainein, metienai, meterkhesthai*
mule: *hêmionos*
multiple: *pleiôn, pleonakis, pollakis*
multiplied by a finite number: *peperasmenos*
multitude: *plêthos*
music: *mousikê*
musical: *mousikos*
must be taken as an enclitic: *enkliteon*
must be taken to be a proparoxytone: *proparoxutonêteon*

name (v.): *kalein, onomazein*
name (n.): *onoma*
natural: *kata phusin*
natural scientist: *phusikos*
naturally: *kata phusin, pephukos* (+ inf.)
nature: *phusis*
necessarily: *pantôs*
need: *deisthai, opheilein, opheilon einai, thelein*
negation: *apophasis*
negative: *apophatikos*
neighing: *khremestikos*
new: *prosiôn*
next: *ephexês*
no doubt: *pantôs*
nominative: *euthus*
noodling: *teretismata*
nose: *rhines*
not a bad idea: *ouden kheiron*
not at all: *mê pantôs*
not possible: *adunatos*
note: *sêmeioun*
notion: *ennoia*
now: *nun*
number: *arithmos*

object: *enistanai*
object of opinion: *doxastikos*
object of perception: *aisthêtos*
object of scientific understanding: *epistêtos*
objection: *enstasis*
obtain: *lambanein*
obtain evidence: *tekmairesthai*
obvious: *prophanês*
occur: *gignesthai*
odd: *perittos*
of a different sort: *heteroeidês*
of arithmetic: *arithmetikos*
of geometry: *geômetrikos*
of horn: *keratinos*

of music: *mousikos*
of physics: *phusikos*
of sight: *horatikê*
of smaller extent: *ep' elatton*
of that sort: *toioutos*
of the sun: *hêliakos*
of wider extent: *epi pleon*
offer: *epangellein*
often: *pollakis*
omens in the sky: *diosêmeia*
omission: *paraleipsis*
on behalf of: *huper*
on theology: *theologikos*
one: *heis, monas*
one must be careful: *eulabêteon*
one must put the circumflex on the penultimate: *properispasteon*
one point at a time: *kata meros*
one's eye is on: *ho skopos teinei*
opinion: *doxa*
opposite: *antikeimenon, enantios*
optics: *optikê*
order: *suntaxis, taxis, thesis*
orderly layout: *taxis*
otherwise: *kath' heteron tropon*
outline: *diagramma*
outnumber: *pleonazein*
outside of: *ektos*
own: *oikeios*

pale: *leukos*
paleness: *leukos*
paradoxical: *paradoxos*
parallelogram: *parallêlogramma*
part: *meros, morion*
partial: *epi merous*
partially: *epi merous*
particular: *kata meros, merikos*
pass over: *parienai*
passage: *khôrion*
passing in front: *hupodromê*
passion: *paskhein*
path: *hodos*
penultimate: *parateleutos*
per se: *kath' hauto*
perceive: *aisthanesthai*
perceptible: *aisthêtos*
perception: *aisthêsis*
perceptive: *aesthêtikos*
perfection: *teleiotês, teleiôsis*
perhaps: *tukhon*
perimeter: *perimetros*
perish: *phtheiresthai*
perish together: *sumphtheiresthai*
perishable: *phthartos*
perishing: *phthora*
permit: *parienai*
perpendicularly: *kata katheton*
persist: *diamenein*
persuasive: *pithanos*
persuasiveness: *pithanotês*
phase (of the moon): *phôtismos*
philosopher: *philosophos*
philosophical: *philosophos*
philosophise: *philosophein*
philosophy of nature: *phusiologia*
physical: *phusikos*
physics: *phusika, phusikê*
Physics: *Phusikê*
piece together additional ...: *episuntithenai*
pitch: *pissa*
place (v.): *paratithenai, tithenai*
place (n.): *topos*
plan: *boulê*
plane: *epipedon, stereon*
please: *boulesthai*
pleasure: *hêdonê*
plenitude: *plêrôma*
pleonasm: *parallêlon*
pleonastically: *ek parallêlou*
point: *sêmeion, stigmê*
point out: *sêmeioun*
poor man: *penês*
posit: *hupotithenai, tithenai*
positing: *thesis*
position: *hupothesis, thesis*
possession: *hexis*
possible: *dunatos*
possibly: *endekhomenôs*
posterior: *husteros*
potentially: *dunamei*
pour: *khein*
pourable: *khutos*
power: *dunamis*
powerful: *karteros*
practice: *epitêdeuein*
precede: *proêgeisthai*
precise: *akribês*
precisely: *hoper*
precision: *akribeia*
predicate (v.): *katêgorein*
predicate (n.): *katêgoria, katêgoroumenon*
predication: *apophansis, katêgoria*
pre-eminently clear: *prodêlos*
preliminary fingering: *prodiapsêlaphêma*
premise: *protasis*
present: *nun, prokeimenos*

present (v.): *tithenai*
present a persuasive argument: *epikheirein*
present with: *paradidonai*
preserve: *phulattein*
prevent: *kôluein*
primarily: *prôtos, prôtôs*
primary: *prôtos, prôtistos*
primitively: *prôtôs*
principle: *arkhê*
prior: *proteros*
privation: *sterêsis*
privative: *sterêtikos*
probative: *deiktikos*
problem: *problêma*
proceed: *proeinai*
proceed along with: *sumproerkhesthai*
proclaim: *boan*
prodeduction: *prosullogismos*
produce: *poiein*
product of craft: *tekhnêton*
progress: *khôrein*
proof: *deixis*
proof from signs: *tekmêriôdês deixis*
proper: *idios, oikeios*
property: *idion*
proportional: *analogos*
propose: *protithenai*
prove: *deiknusthai*
provide: *paradidonai*
providence: *pronoia*
proximate: *prosekhês, prosekhôs*
prudence: *phronêsis*
public speaker: *rhêtôr*
pure: *eilikrinês, katharos*
purely: *eilikrinôs*
purpose: *khreia*
pursue: *meterkhesthai*
put forward: *proagein*
puzzle: *aporia*
puzzle about: *aporein*

quality: *poion, poiotês*
quantity: *poson*
quarter tone: *diesis*
quod erat demonstrandum: *hoper edei deixai*

rain: *ombros*
raise a puzzle: *aporein*
raise a question: *aporein*
raise objections: *enistanai*
rational: *logikos*
ray of vision: *opsis*
reach: *phthasai* (+ *eis*), *aphikneisthai, hêkein, katantan*
read: *anaginôskein*
reason: *aitia, aition, logos*
reasoning: *logismos*
recall: *anamimnêskesthai*
receive: *hupodekhesthai*
recognise: *gnôrizein*
rectilinear: *euthugrammos*
reduction: *apagôgê*
refer to: *dêloun, kalein, legein, onomazein*
refutation: *elenkhos*
refute: *elenkhein*
relate: *ekhein, historein*
relate to: *ekhein pros*
relation: *pros ti*
relative: *pros ti*
relative position: *logos*
remain: *leipesthai, menein*
remove: *khôrizein*
render: *apodidonai, poiein*
repetition: *epanalêpsis*
reply: *apokrinesthai*
reputable: *endoxos*
require: *axioun*
respond: *phanai*
response: *apantêsis*
responsible for: *aitios*
rest: *êremia*
result: *sumbainein*
return: *epanienai*
reversed: *huperbatos*
rhetorically: *rhêtorikôs*
rich man: *plousios*
ridiculous: *katagelastos*
right angle: *orthê*
right now: *en tôi paronti*
right side: *dexios*
rise: *anatellein*
role: *khreia*
run through: *diêkein*

say: *eipein, erein, legein, phanai, phaskein*
science: *epistêmê*
scientific: *epistêmonikos*
scientific understanding: *epistêmê*
sea: *thalattios*
second: *deuteros*
see: *skopein, suneidein*
see within: *entheôrein*
seed: *sperma*
seek: *zêtein*

seem: *dokein, phainesthai*
segment: *tmêma*
self-evident: *autopistos*
self-moving: *autokinêtos*
self-subsistent: *authupostatos*
semicircle: *hêmikuklion*
sensation: *aisthêsis*
sense: *dianoia*
sensible: *aisthêtos*
separable from: *khôris*
separate (adj.): *khôristos*
separate (v.): *khôrizein*
separately: *idiai, khôris*
series of predication: *sustoikhia*
serve as subject: *hupokeisthai*
set out: *ektithenai*
several: *pleiôn*
shadow: *skia*
shape (n.): *skhêma*
shape (v.): *skhêmatizein*
shine: *phôtizein*
show: *deiknusthai, epideiknusthai*
sickness: *nosos*
side: *meros, pleura*
sight: *opsis*
signify: *sêmainein*
simple: *haplous*
simpliciter: *haplôs*
simply: *haplôs*
single: *heis*
sit down: *kathêsthai*
situation: *skhesis*
skilfully: *tekhnikôs*
snub: *simos*
snubness: *simotês*
so and so: *deina, toios*
so-called: *kaloumenos, legomenos*
solution: *epilusis, lusis*
solve: *epiluein*
some: *meros*
someone who is initiated with someone else: *summustês*
something in common: *koinônia*
something that instils: *empoiêtikos*
sometimes: *pote*
song: *melos*
soul: *psukhê*
sound: *apêkhêsis, phthongos, psophos*
sounding: *apêkhêsis*
speak: *eipein, erein, legein, phanai*
species: *eidos*
specific: *eidikos*
specifically: *idiôs*
specify: *prostithesthai*
sphere: *sphaira*
spherical: *sphairikos*
sphericity: *sphairoeidês*
Spherics: *sphairika*
spontaneous: *automatos*
spot: *stokhazesthai*
spread: *diadidonai*
square with: *sunaidein*
standing before: *epiprosthesis*
standstill: *epistasis*
start (n.): *arkhê*
start (v.): *arkhesthai*
state: *eipein, erein, legein*
state of being motionless: *akinêsia*
statue: *andrias*
status: *logos, skesis*
stay: *menein*
stellar: *astrôios*
stereometry: *stereometria*
stick: *xulon*
stimulate the juices: *hugrainein*
stone: *lithos*
stone cutting: *laxeutikê*
stop: *histanai, pauesthai*
straight: *euthus*
strictly speaking: *haplôs*
string: *khordê*
study: *episkeptesthai, theôrein, theôria poiein*
subject: *hupokeimenon*
subordinate: *hupallêlos*
subsist: *huphistanai*
subsistence: *hupostasis, huparxis*
subspecies: *eidos* (+ *hupo*)
substance: *ousia*
substantial: *kat' ousian*
successive: *ephexês*
successively: *ephexês*
such: *toios, toioutos*
such as cannot be traversed: *adiexitêtos*
suffice: *arkein*
summit: *akrotaton*
sun: *hêlios*
superior: *kreittôn*
supply in thought: *prosupakouein*
suppose: *huponoein*
surely: *pantôs*
surface: *epiphaneia*
subsistent: *huphestêkos*
sweet: *glukus*
swell up: *ônkousthai*
sword: *ksiphos*
syllogism: *sullogismos*
symbolically: *sumbolikôs*

take: *hairein, lambanein, paralambanein*
take additional: *proslambanein*
take note of: *mê lanthanein*
take up: *analambanein, lambanein*
talk: *dialegesthai*
talk about: *phanai*
talk to: *entunkhanein*
task: *ergon*
teach: *paradidonai*
tend to: *therapeuein*
tendance: *therapeia*
term: *horos, onoma, phônê*
terminal: *teleutaios*
terminate: *teleutan*
terrestrial: *khersaios*
testing: *dokimasia*
text: *lexis, rhêton*
that follows: *ephexês*
that succeeds: *ephexês*
the departed: *hoi katoikhomenoi*
the effect that: *toioutos*
the like: *toioutos*
the masses: *hoi polloi*
the number of ways in which: *posakhôs*
the qualified: *poion*
the sort of thing: *toioutos*
the state of not having considered something: *anepistasia*
the subject: *ta pragmata*
the task set: *prokeimenos*
the view that there is no apprehension: *akatalêpsia*
theology: *theologia*
theorem: *theôrêma*
theory: *theôria*
thereupon: *ephexês, euthus*
thesis: *problêma*
thicken: *puknoun*
thicken down: *katapuknoun*
thing: *phusis, pragma*
thing that is done: *prakton*
think: *nomizein, oiesthai*
think about: *phrontizein, skeptesthai*
think of: *noein, epinoein*
this kind of: toios
this particular thing: *tode ti*
this sort of thing: *toioutos*
this way: *toiôs*
thoroughly: *panu*
thoroughly examine: *exetazein*
thought: *dianoia, epinoia*
three cubits tall: *tripêkhus*
through the continuum: *kata sunekheian*
through the impossible: *dia tou adunatou*
timber: *sanis*
time: *khronos, pote*
together: *hama*
tomorrow: *aurion*
tool: *organon*
topic: *theôrêma*
total: *pas*
transcend: *epanabainein, exairesthai*
transpose: *metatithenai*
transpose upwards: *huperbibazein*
travel: *hodeuein*
traverse: *dierkhesthai, diexerkhesthai*
treatise: *pragmateia*
triangle: *trigônon*
true: *alêtheis*
truth: *alêtheia*
truthfully: *kat' alêtheian*
turn back: *anatrekhein*
twenty times: *eikosaplasion*
twice as many: *diplasios*
twofold: *dikha*
two-footed: *dipous*

ugly: *aiskhros*
uncertain: *abebaios*
unclear: *amudros*
unconvincing: *apithanos*
undergo: *paskhein*
underlie: *hupoballesthai, hupokeisthai*
underlying subject matter: *hupokeimenon*
understand: *akouein, epistasthai, noein, prosupakouein*
unequal: *anisos*
unit: *monas*
unitary: *heis*
unity: *heis*
universal: *katholikos, katholou*
universally: *katholou*
universally speaking: *katholou*
unjust: *adikos*
unknowable: *agnôstos*
unnatural: *para phusin*
unnaturally: *para phusin*
unqualified: *haplôs*
up to what point: *mekhri posou*
upwards: *anô*
usage: *khrêsis*
use: *paralambanein*
useful: *khrêsimos*
utmost: *eskhatos*
utterly refute: *dielenkhein*

valid: *hugiês*

valuable: *timion*
vegetables: *lakhana*
very: *panu*
very last: *teleiôtikos*
very many: *pleistoi*
virtue: *aretê*
void: *lumainesthai*

wage war: *polemein*
walk: *badizein*
walking upright: *orthoperipatikos*
want: *boulesthai, thelein, zêtein*
war: *polemos*
way: *tropos*
way downwards: *kathodos*
weakness: *astheneia*
weight: *baros*
what connects: *sunagôgos*
what has been proposed: *prokeimenos*
what is evident: *enargeia*
what it is: *to ti esti*
what sort of thing: *poion*
whatsoever: *haplôs*
which kind of: *poion*
which transcends: *huper*
white: *leukos*
white lead: *psimmuthios*
whiteness: *leukotês*
whole (n.): *holon*
whole (adj.): *pas*
whole tone: *epogdoos*
wine: *oinos*
winged: *pterôtos*
winter: *kheimôn*
wisdom: *sophia*
with a head: *kephalôtos*
with a rudder: *pedaliôtos*
with good reason: *eulogôs*
with position: *thetos*
with the character of a principle: *arkhoeidês*
within: *entos*
without: *dikha, khôris*
without demonstration: *anapodeiktôs*
without position: *athetos*
without qualification: *haplôs*
wooden: *xulinos*
word: *logos, phônê*
work (n.): *pragmateia*
work through: *gumnazein*
worthy of conviction: *pistos*
worthy of discussion: *axiologos*
write at the start: *arkhên poieisthai*

Greek-English Index

Index of Names

References are to the page and line numbers in the margins of the translation.

Subject Index

References are to the page and note numbers of this book.

Lightning Source UK Ltd.
Milton Keynes UK
UKOW05f0610250414

230576UK00001B/51/P

9 781472 557995